Foreword

The study of cultural ecology is essentially the investigation of human adaptability. The roots of human adaptive behavior lie deep in man's phylogenetic history. Hans Kummer illuminates this by his careful analysis of primate social life, showing how the genetically programmed behavior of baboons and other monkeys is subject to adaptive modification to meet the exigencies of both their physical and their social environment.

Dr. Kummer does not involve himself in the fruitless argument whether these infrahuman animals have "culture" in the anthropological sense. He is fully aware that they do not have the elaborately coded system of symbols which is the essence of the human context and the very stuff of culture. What he does show, however, is that monkey bands show patterned forms of behavior that are adaptive to local situations. Among these situations are those created by special elements in the physical environment, such as food resources and sleeping areas; and these affect the nature of collaborative action. Collaborative action, in turn, requires the structuring of social relationships among the primates, necessitating a further adaptive modification of behavior. At the same time, limits on this adaptive capacity in each species of animal are set by its genetic preprogramming.

The analysis of primate ecological adaptation is based primarily upon field studies by Kummer and his colleagues. Dr. Kummer has made ingenious use of natural experimental situations to discover the nature of primate adaptability. Thus he has examined the social organization of a single species under diverse environments and has studied the behavior of different species in a constant environment. These field studies have been enriched by equally ingenious experiments. He also brings into consideration relevant research by other students of primate ethology.

Although this book does not deal with humans nor does it (in a strict sense) deal with culture, I am glad to include it in *Worlds of Man,* which is a series of books on human cultural ecology, because by traversing these boundaries we perceive the crucial region lying between the cultural and the noncultural. We are thus able to see both the limitations placed on behavior by inherited characteristics and the scope of behavioral adaptation.

This book has a broader mission as well: it is a corrective to the recent spate of popular works that have endeavored to extrapolate from animal behavior to man. I have in mind such simple-minded conceptions as that man is the inheritor of a territorial imperative, as Robert Ardrey has argued, or an aggressive instinct, as Konrad Lorenz implies. What implications can really be drawn for the biological programming of man from the simple lives of stickleback fish or herring gulls, when man's closest relatives display such complex and varied behavior? Dr. Kummer demonstrates that in crucial matters relating to social organization, the various species of primates have diverse repertories of innate behavior and these, in turn, they modify to meet the exigencies with which their environment confronts them. Dr. Kummer shows clearly that ecological adaptation, both biological and cultural, is a complex phenomenon, and thus points up the inadequacies of popular oversimplifications. Without trying to extrapolate from primate to human behavior, but rather by examining the forces that shape that behavior, he succeeds in giving us real insight into the world of man.

Walter Goldschmidt

Primate Societies

Group Techniques of Ecological Adaptation

HANS KUMMER

AldineTransaction
A Division of Transaction Publishers
New Brunswick (U.S.A.) and London (U.K.)

First paperback printing 2007
Copyright © 1971 Hans Kummer.

All rights reserved under International and Pan-American Copyright Conventions. No part of this book may be reproduced or transmitted in any form or by any means, electronic or mechanical, including photocopy, recording, or any information storage and retrieval system, without prior permission in writing from the publisher. All inquiries should be addressed to AldineTransaction, A Division of Transaction Publishers, Rutgers—The State University, 35 Berrue Circle, Piscataway, New Jersey 08854-8042. www.transactionpub.com

This book is printed on acid-free paper that meets the American National Standard for Permanence of Paper for Printed Library Materials.

Library of Congress Catalog Number: 2006051023
ISBN: 0-202-30904-5
    978-0-202-30904-0
Printed in the United States of America

Library of Congress Cataloging-in-Publication Data

Kummer, Hans, 1930-
  Primate societies : group techniques of ecological adaptation / Hans Kummer.
    p. cm.
  Originally published: Chicago : Aldine-Atherton, 1971.
  ISBN 0-202-30904-5
  1. Primates—Behavior. 2. Social behavior in animals. I. Title.

QL737.P93K79   2006
599.8'156—dc22   2006051023

Contents

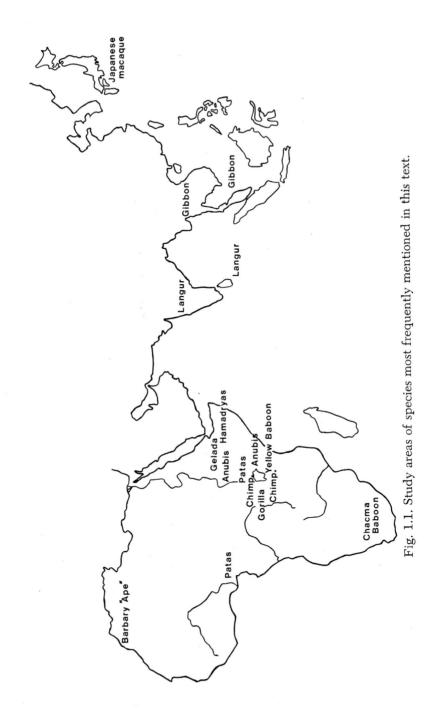

Fig. 1.1. Study areas of species most frequently mentioned in this text.

Chapter 1

"CULTURE" AND THE CONCEPTUAL FRAME OF BIOLOGY

In this volume, the anthropological concept of cultural ecology is subjected to a double stress. First and worse, it is delivered into the hands of a disciple of an alien science, a zoologist, who approaches it with a frame of thought that has virtually no place for culture as this term is usually understood. Second, it is applied to tribes that, though close relatives to man, are nevertheless nonhuman animals. These are the costs of our comparative outlook; its returns will have to be judged by the reader.

The fact that we shall deal with nonhumans is a difficulty that, I hope, will be overcome by the first and last chapters of this book, where I shall try to introduce the reader to primate societies with as much of an anthropologist's outlook as I can muster. The difficulty of the alien frame of thought, however, is not so easily overcome. The conceptual world of the zoologist is as much part of this text as the data presented in it and must therefore be made explicit. Explicitness is commendable for yet another reason: When thinking in terms of ecological adaptiveness, anthropologists use a zoological concept outside its native context of thought. Our first task here is to confront the student of anthropology with the context of zoological thinking from which the term "adaptiveness" is taken.

The main concepts that guide biological observations, experiments, and discussions can be grouped into five dimensions or viewpoints. First, there is *structure*. The structural or, as it were, the anatomical, outlook describes momentary situations of a living system; it records the proportions of a bone or the composition of a group. Such situations, however, are constantly changing, forcing us to think in terms of processes, such as the growth of a bone or the division of a group. Biological processes are traditionally judged from two contrasting viewpoints that form the second and third of our dimensions: Processes leading to the situation in which we are interested are analyzed as possible *causes* of that situation. Processes emerging from the reference situation are its possible *functions*. By "function" we mean the effect of a process on the success of the living system in which the process takes place. Thus an "adaptive function" enhances the survival chance of the animal or population in which the process takes place.

The fourth and fifth dimensions of biological thinking deal with a larger time scale. *Ontogeny* is the process by which a fertilized egg cell, endowed with a set of genes, develops into a mature and finally an old adult. The study of ontogenetical life cycles attempts to untangle the enormously complex internal processes of individual development. It also analyzes the inputs from the social and ecological environment that affect the course of development. The directing action of such external, nongenetical stimuli is "modification" in biological jargon. Finally, biologists are interested in the long-term processes that alter the genetic endowment from which ontogenies start. These processes are summarized under the heading of *evolution*.

This, in rather crude form, is the biologist's world of concepts into which he fits his observations. Like all such frameworks or viewpoints, biological thinking is useful only within limits; it simplifies phenomena in which it is not primarily interested. The phenomenon of "culture," for example, can readily be recognized as a "social modification." However, this biological definition ignores some of culture's most important aspects. A biologist is helpless in the face of such concepts as "attitudes" or "value system," not because he denies their reality, but simply because he has no research tool for detecting any such thing in an animal. He cannot interview his subjects and thus never obtains an inside view of an animal society. He can only read behavior.

For a biologist, the term "culture" comes to mean a set of behavior characterized by its *origin*. An individual develops a particular behavior partly because its genetic endowment directs its development, and partly because the environment feeds information into the process of development. If a behavior were entirely programmed by genes, it could be called "innate"; if it were an exclusive product of environmental stimuli during the animal's ontogeny, it could loosely be called "acquired" behavior. In reality, these extremes do not occur. Although the swimming movements of a fish and the skill of a translator come close to the extreme forms, the fish will never swim unless it finds suitable conditions for its development, and the translator's faculties depend on a genetic basis that is uniquely human. Each observed trait is thus shaped by both information contained already in the egg cell and by information drawn from the ontogenetical environment; nevertheless, the distinction between the two sources of information is real. If two fish with diverse genotypes, but raised under the same environmental conditions, develop different swimming movements, it is safe to say that the *difference* is an effect of genotypes, not of environment. If, on the other hand, identical twins acquire different languages when raised in different nations, the environment must be responsible. The important thing to note is that only a difference between traits, not a trait as such, can be called "innate" or "acquired."

This puzzling statement, which is the solution of the now obsolete nature-nurture controversy, may need some thinking. The argument is that no trait can possibly develop in total seclusion from either environmental influences or an ever so indirect action of genes. A person speaks French not only because he grew up among Frenchmen, but also because he inherited a genetic basis for language. The trait is neither "acquired" nor "innate" but both. But speaking French *rather than Italian* can be caused by the environment alone; the difference is purely acquired. Or, in an analogy: It takes a drum and a drummer to produce a sound. Nobody would try to differentiate between sounds produced by the drummer and sounds produced by the drum. But we can very well discuss whether two recorded performances sound different because of a new drummer or a new instrument.

With this in mind, we can now approach a distinction whose making I consider one of the important tasks of this book and of research in our field: a clear understanding of the main types of biological adaptations and of their mutual relationship.

The first type is so-called *phylogenetic adaptation*. This is an adaptation of the evolving genotype, not of the ontogenetical process. It occurs when two populations of animals or people have different behavioral adaptations because their egg cells were endowed with different genes. These two populations will, as a rule, develop different forms of behavior even when they are raised in identical environments. Obviously, the process of adaptation occurred before their egg cells were formed, by an evolutionary sorting out of advantageous genotypes. Phylogenetic adaptation is a slow process. It can provide only a generalized behavioral program which is adapted to the *general* features of the habitat in which the interbreeding populations evolved.

The egg cell starts out from this general array of available programs, and at this point, the second type of adaptation takes over. It is *adaptive modification* and is manifested when two populations with the same genotypes develop different behavioral programs in adaptation to the particular environments in which they happen to grow up. The fact that a monkey grows hair is a phylogenetic adaptation, but the fact that he grows thick and long hair when he is exposed to a cold climate is an adaptive modification. Similarily, a baboon may be phylogenetically programmed to spend the night above ground, but his consistent choice of a particular cliff or tree grove is a modification induced by local conditions and by the traditions of his group.

Adaptive modifications can be divided according to the source of these modifying stimuli. If they stem from the physical environment, such as the terrain or the climate, or from other species living in the same habitat, the modification can be termed ecological. However, the individual's behavior can also be modified by its mother or by the group in which it is raised. If such social modification spreads and perpetuates a particular behavioral variant over many generations, then we have "culture" in the broad sense in which a student of animals can use the term. It can be defined as follows: Cultures are behavioral variants induced by social modification, creating individuals who will in turn modify the behavior of others in the same way. If this definition is accepted, the behavior of two groups with the same gene pool and with the same type of habitat can differ only by culture. The definition states nothing about the precise mechanism of the social modification (because it is unknown in most cases), nor about the categories of behavior that should or should not be accepted as cultural (because animals seem to offer no meaningful criteria for such a distinction).

The concept of culture obviously loses a great deal when accommodated to the dimensions of biology. What we can gain from the operation is the wider context of evolution from which culture emerged as one possible way of life, a context from which it can not break loose and which therefore has to be analyzed. Adaptation by culture is only one way of adapting. Its stage is prepared by phylogenetic adaptations that affect cultural developments. In the human species this stage appears so large that its existence and limits are easily forgotten. In the case of nonhuman primates, phylogenetic programming offers much less choice for social modification and thus for rapid change. The investigator's attention is focused on their phylogenetic dispositions and on the problems of distinguishing them from modifications.

The distinction between cultural and noncultural components of behavior is difficult to make, and for most behavioral adaptions in primates it has not even been attempted. In the first part of this book, I must therefore neglect it entirely, describing the ecological functions of primate social behavior in professed ignorance whether such adaptations are cultural, ecological, or phylogenetic in origin. In the second part, however, I shall address myself to these distinctions and to the *process* of adaptation. To know the type to which an adaptation belongs is not merely to gain an academic insight into its origin. The speeds at which the different types of adaptation can occur are so enormously different that to know the origin is to know the prospects of future flexibility.

After explaining the conceptual world from which I must approach the subject, I should add a remark on the material presented here. Although primate societies have been discussed from the viewpoint of their adaptiveness for about ten years now, the factual knowledge on such correlations is meager. Most of the available data are not even quantitative, let alone experimental. Many of the speculations that were printed a few years ago have been badly shaken by more recent information. When, in 1960, Kurt and I found the first example of a one-male group organization in old-world monkeys, this social structure was interpreted as being an adaptation to the extremely harsh semi-desert habitat of the hamadryas baboon which we had studied. In the ten years since then, more and more primate species have been found to live in one-male groups—and most of them are forest monkeys which inhabit the richest habitat that dry land can offer. In a recent review of the correlations between the social structures and the habitats of all investigated African cercopithecine monkeys, the primatologist Struhsacker finds little support for an understanding of social structures as simple correlates of simple classes of habitats.

Solid research has yet to begin, and we shall therefore use an ungraceful amount of speculation. I propose in the following chapters to describe traits of primate societies and a way of thinking about their adaptive function. The results of this thinking, however, should be viewed as hypotheses at best.

One of the reasons for such caution is the concept of adaptiveness itself. To say that a trait is adaptive is, by itself, vague: A few examples will show the possible complexities. In certain human populations of Africa, the recessive gene for sickle-cell anemia is surprisingly high. Up to 45 per cent of the individuals are heterozygous for this allele which, in homozygous subjects, may cause a lethal anemia. There is evidence that heterozygous carriers are more resistant to malaria than genetically "healthy" subjects. The success of the heterozygous condition apparently explains the enormous frequency of the lethal factor in the investigated populations. We may define adaptiveness as the quality of a trait which, under a given range of conditions, increases the number of offspring of the carriers of this trait. (Note the technical acultural content of this biological definition.) If the above conclusion is correct, sickle-cell anemia is an adaptive trait in these populations, even though it may kill.

Male hamadryas baboons have an inhibition which prevents them from appropriating females belonging to other males of their troop. A poorly developed inhibition should allow a deviant male to collect the females of subordinate troop members; he would thus produce more offspring than his inhibited rivals. A low-level inhibition appears "adaptive" for its carrier, but it is likely to be maladaptive in its effect on the social stability of the troop.

Some ungulates chew with stereotyped motor patterns of the jaw. In camels, the mandible alternates between a motion to the right and a motion to the left, whereas duikers ruminate on one side for quite a while and then shift to a similar series of motions on the other side. The adaptiveness of these patterns does not lie in their particular form, but in their rigidity as such, which prevents the formation of chewing habits that would wear only the teeth on one side. Adaptive function must be sought on the appropriate level.

A primate male may have a stronger than usual tendency

to approach and distract predators. As long as only one or two males of a group are thus inclined, the trait may be called adaptive defense of the group, but the same trait will assume a negative value if too many males of the group expose themselves to the danger of being killed.

Chimpanzees can paint. While it is difficult to imagine the survival value of such artistry, it is possible that the performance is an output of a behavioral subsystem that is part of a larger, adaptive system.

A conclusive statement on the adaptiveness of a trait would require data on its positive and negative effects on many levels of organismic and social organization, and under a wide variety of environmental situations. This volume can offer no such data. Every one of its statements on adaptation would in principle require experimental testing. Since we cannot reasonably hope to carry out such experiments on the appropriate scale, I shall try, in Chapter 5, to outline some correlative methods that can improve the quality of our present knowledge.

SUMMARY

1. The main dimensions of biological thinking are structure, causation, function, ontogeny, and evolution.

2. Phylogenetic adaptation is an adaptive change of the gene pool by mutation and selection; adaptive modification is the shaping of the ontogenetical process by the individual environment.

3. In the limited conceptual framework of biology, the term "culture" can be defined only as a behavioral modification induced by the social environment.

4. A given trait can be adaptive in one functional context or level and maladaptive in others.

Chapter 2

AN INTRODUCTION TO
PRIMATE SOCIETIES

A SOCIETY OF HAMADRYAS BABOONS

To provide the reader with a feeling of primate social life, I shall describe a society familiar to me, and I shall describe it, not in theory, but in the way it presents itself on an average day.

It is a troop of hamadryas baboons that inhabits the arid grassland, interspersed with thorny acacia trees and bush, in rolling open country at the southern edge of the Danakil desert in Ethiopia (Fig. 2.1). It is dawn. About one hundred baboons are scattered over the narrow ledges of a white vertical cliff standing above the sandy bed of a presently dry river (Fig. 2.2). Most baboons are still asleep; facing the cliff, they quietly sit, holding onto knobs and fissures of the rock with their hands. They have passed the whole night in these uncomfortable postures, although they woke up frequently to change position or to raise a chorus of grunts at some disturbance away in the darkness. They are diurnal animals, and they protect themselves against nightly attacks from predators by holding onto this rock. Nevertheless, a

Fig. 2.1. An arid hamadryas habitat in Eastern Ethiopia with thornbush and sleeping cliffs.

Fig. 2.2. Sleeping cliffs of a hamadryas troop, about 20 yards high. Black spots mark preferred sleeping ledges.

leopard may kill one or two troop members who sit too close to the foot of the cliff.

As the day approaches, baboons wake up here and there, scratch or shake themselves, then lower their heads for another doze (Fig. 2.3). At sunrise, the animals gradually rise to all fours. Small parties of baboons, still stiff from the night's cold, slowly climb along the ledges toward some vertical fissure that leads them to the top of the cliff and into the sunlight. More often than not, such a party consists of one adult grey-mantled male, a number of the much smaller, brown-haired females, and a few infants. These are the one-male groups, the smallest social units of the troop, usually about five animals.

On the gravel slope above the cliff, the baboons sit down in the sun as if to warm themselves. On cool and cloudy mornings, they huddle together in small, tightly-packed clusters, usually one-male groups. Very often, a subadult male sits near the group, separated from it by a few feet. These are the so-called followers, half-grown males that have attached themselves to a particular one-male group. They keep their distance because the group male is somewhat intolerant of their touching the females.

During the next one or two hours, the baboons remain on the sunny slope for a social session. Females begin to groom the mantle of their male, who leisurely adapts his posture to their activity. A female in oestrus presents her perineal swelling to her male, who may then mount and copulate with her. Other females sit a few yards away, grooming their infants or one another, or engaging in a grooming session with the follower of their group—an activity that may be cut short by a stare from the group male. The follower then escapes, and the female, by way of appeasement, may run to present her rump to her male even if she is not sexually receptive. Grooming with the follower is the utmost that the group male tolerates in his females. They are not permitted to copulate with the follower, although they sometimes succeed in doing so after deliberately walking to a rock that hides them

Fig. 2.3. A hamadryas male and his females sitting on their customary sleeping ledge. The male is threatening a neighboring family group.

from the group male's view. If the pair are caught in the act, the group male attacks his female and bites her on the nape of the neck, while the follower is allowed to slip away.

Among the grooming one-male groups one detects other types of social units. An old adult male may sit or groom with a big subadult male. Other adult males simply sit alone. About 20 per cent of the adult males in the troop are singles that have no females of their own. They never approach or interact with any female, but rather associate with each other for a few minutes. Nevertheless, they are troop members like the group males.

Older infants and juveniles, mainly males, have by now assembled into several play groups. The young females their age have remained with their mothers and joined the grooming sessions of the adult females. Other juvenile females cannot join a play group because they are already full members of a nascent one-male group. These "initial groups" (Fig. 2.4) consist of a young barely adult male and a single juvenile female. The male has "kidnapped" her from her maternal group and now forces her to follow him by staring at her or chasing her whenever she is more than a few steps away from him. Although she is still far from puberty, he herds her and prevents her from contacting other troop members. The conditioning process may take several days of intermittent escapes by the young female and chases by her male. Such scenes are rare in mature one-male groups, because the adult female has long ago learned that any interaction by her with a troop member outside her own group or any attempt to leave the group male will provoke him to attack her. She therefore follows her male up from the sleeping-ledge as if she chose to do so (Fig. 2.5).

In the resting troop, the various one-male groups have so far appeared to take little notice of each other—except for the youngsters, who may move freely throughout the troop. A female may have inspected a new-born infant in a neighboring group, only to rejoin her own group hastily. Once in a while two adult males have chased each other back and

Fig. 2.4. Juvenile female of an initial unit grooms her male in the early sun.

Fig. 2.5. An adult male preceding his two females. He frequently looks back and checks on their following response.

forth over the rocks for no apparent reason and without any physical contact or injury. In fact, the adult members of a group refrain from interacting with adults of other one-male groups, although they sit in full view of each other and only a few feet apart.

At around eight or nine in the morning, the occasional shifts of groups on the slope become more frequent. Here and there, an adult male scratches, gets up, and moves toward the periphery of the troop to sit down there, his females following him. If one carefully maps these shifts, it becomes apparent that the males pay much more attention to each other than superficial inspection indicates. The shifts of neighboring males are highly interdependent. If a particular old male moves in a certain direction, the surrounding males, after a minute or two, turn on their buttocks to face the direction in which he has gone, or execute a parallel shift. A young male may make a similar move without obtaining any such response from his neighbors and he may soon after return to his former place.

During this process, the males are as silent as before, but their obsessive scratching before nearly every change of place betrays the conflicts to which the multiple interaction subjects them. The process, in fact, is an important decision process about the direction of the daily foraging trip for which the troop now prepares. This is almost exclusively an adult males' affair. The group males, who never groom one another, now begin to interact openly before or after certain shifts. Often a male gets up, walks over to his neighbor hesitantly, presents his brilliant red buttocks in a rapid turn, and then hastily retreats again as if he had approached his neighbor too closely—which he did, if compared with the usual distance between group males. Contextual analysis suggests that this peculiar presentation probably functions as a notification to a neighbor of one's imminent departure, a signal that seems to say, "Watch out, I'm leaving now—in case you want to follow."

As a consequence of the ever more frequent shifts, the

troop changes its shape like an ameba. Nobody has departed yet, but here and there the troop's periphery protrudes in a kind of pseudopod which may persist or withdraw again. After perhaps half an hour, a number of male baboons in the center of the troop finally get up in quick succession and walk toward one of the pseudopods. Their straightforward walk is quite different from the previous hesitant shifts at the periphery; many animals begin to move in parallel direction, and suddenly the troop is on the move. It is reasonable to assume that the pseudopods were proposals for particular directions by peripheral males, and that some influential male near the center finally made a decision. On another morning, a different male may initiate the departure. The troop, which now leaves the cliff rapidly and in a dense column, is clearly not led by any one particular animal. The baboons at the head are constantly replaced, and it looks as if everyone knew the direction of the march. The juvenile play groups now dissolve and the players rejoin their maternal groups. Young infants are carried at their mother's belly, older ones on the back.

The troop continues its rapid procession for about half an hour; it holds a steady course from which it deviates mainly in order to use open ridges or river beds (Fig. 2.6). Then, the walking slows down and the column begins to split up; the baboons begin to forage. In the hours before noon, the troop has split into several large subunits that are no longer in sight of each other. These subunits, the so-called bands, are constant in membership; they did not appear as distinct units at the sleeping-cliff, but now they tend to separate. In the process of foraging, however, even the bands are often sufficiently scattered for the one-male groups to emerge as spatially distinct units (Fig. 2.7). A male and his group typically pick the same acacia tree.

Hamadryas are mainly vegetarians. The flowers, young leaves, and beans of the many species of acacias are their most important food source in the southern Danakil plain. They pick them with one hand and rapidly stick them into

Fig. 2.6. A troop after departing from its cliff. One-male groups are not evident during this rapid initial progression.

Fig. 2.7. A one-male group resting at mid-morning after splitting from the troop.

their mouths one by one. At the beginning of the rainy season, they feed on young grass leaves in similar fashion, and later in the year, they strip the grass ears by pulling them through their closed jaws with one hand.

In the dry season, they must turn to tougher food, such as the bitter leaf bases of wild sisal, the leathery leaves of the Dobera tree, and roots. Large food items are almost nonexistent in the Danakil plain; the tiny bits are always eaten on the spot and offer little cause for competition. Insects are caught with one hand. Once, two of our baboons were seen to carry and quarrel about a half-eaten young dik-dik that one of them had killed. Normally, however, each baboon feeds by himself on the spot and does not share or compete for pieces of food. An infant will sometimes run over to its mother, watch her closely while she is eating, and then sniff her mouth. Thus, young baboons may learn what food their mothers select.

After a long arid stretch that was passed by scattered one-male groups while feeding here and there, a band of the troop assembles again in a dense grove of acacia trees on a river bed. They spend an hour or so in the trees, feeding and resting. Suddenly, several adult males emerge from the undergrowth, all watching a large old male of their band as he treks on down the dry river bed with his females. Many of the band begin to follow and even pass him. Soon he is walking at the rear of the column, which continues traveling in the direction that he indicated. After two miles' walk, the band arrives at a rocky barrier that crosses the dry river bed. At the foot of this cliff is a pond of open water (Fig. 2.8). Another band of the troop is already resting in the shade of the meager gallery forest. The newcomers also sit down.

It is 1 p.m. The first part of the daily trip has been successful. Although the troop has already traveled more than five miles, most of it under the hot sun, it has touched on a sufficiently well-stocked grove of acacias and it has reached water. Most likely, recollections of these and other important spots in the troop's home range governed the decision process of the troop males before departure.

The baboons rest for half an hour, then they go down to the water to drink, a few at a time. Several baboons dig shallow holes in the sand near the edge of the pond; they wait for a while, then drink the clear, filtered water that rises, instead of the greenish liquid in the pond. Under the trees, play groups are again formed; the adults groom.

At 2 p.m. the bands depart, mostly without a further decision process. Feeding as they walk, they arrive at the sleeping-cliff in the late afternoon. As the first band approaches in a long column, an adult male at its head suddenly looks back, stops, stands on his hind legs, and excitedly inspects the arriving animals. One of his females is missing. He spots her among the last of the arriving baboons, dashes past the column, and violently bites her on the back (Fig. 2.9). She screams and urinates but follows him closely as he hastens up to where his other females are waiting. The bands settle for another social session about the sleeping-cliff. At dusk, the one-male groups will seek their usual ledges in the cliff.

One of the bands that departed with the others in the morning is now missing. It left the waterhole in a different direction and now approaches the next available cliff, Red Rock, situated several miles from the White Cliff where it spent the previous night. As the band ascends the last ridge and the Red Rock comes into view, it is greeted by resounding, deep, double barks from the Rock. About 60 baboons are already sitting on the grassy slope below Red Rock. The newcomers immediately sit down and look over at the Rock, but do not answer the barks. After an uneasy waiting period, a few males of our band hesitantly begin to approach the Rock and its residents. After a while, the band slowly follows and enters the Rock from the side opposite to the residents, with whom they will form tonight's troop.

Not all bands of an area are on troop-forming terms with each other. Had the first-comers at Red Rock been somewhat less familiar, our band would probably have withdrawn and made for the White Cliff. The troop is an unstable sleeping

Fig. 2.8. Water hole in a river bed at the beginning of the dry season.

Fig. 2.9. A male attacks one of his females after she has left him to drink from the river by herself.

community of several bands that are sufficiently familiar with each other to share the same cliff.

The reluctance of hamadryas baboons to approach unfamiliar bands is well justified. When, on one exceptional evening, a band from Red Rock attempted to spend the night on the White Cliff, two males got into a fight. Unlike ordinary fights within a well-integrated troop, this one immediately developed into a violent, large-scale battle that involved many males of both the resident and the intruding bands (Fig. 6.2). Significantly, the males did not fight over sleeping space, but over females.

Female-possession among hamadryas baboons ordinarily is respected within the band and within a well-integrated troop. The usually peaceful life within the troop is the effect of specific male inhibitions which prevent them from touching each others' females, as will be described later. These inhibitions, apparently, are responsible for the careful spacing and the lack of interaction between the males of one-male groups that we noted in the morning. There are apparently no effective inhibitions against encroaching on females of another troop, but intertroop conflicts are nevertheless very rare because the troops keep apart.

GENERAL CHARACTERISTICS OF
NONHUMAN PRIMATE SOCIETIES

The concrete background of hamadryas society will permit us to visualize the general qualities of primate social life that most strikingly contrast with human societies.

Communication

Primates, lacking symbolic language, can communicate only about the here and now. Their vocalizations, gestures, and facial expressions *inform* the recipients of what could be called the present "mood" of the actor, about what he is likely to do next. A threatening stare indicates that he is likely to attack in the next few seconds, but it communicates nothing about what he will be doing an hour or a day from now—except that, under the same circumstances, he will probably stare again. The same is true for communication about space. A monkey can indicate that he claims access to a stand of mushrooms which both he and his partner can see at the moment, but he can communicate nothing about the location of the same stand when it is behind the next ridge, except by leading the other to the spot. Hamadryas males preparing for departure are able to signal the direction in which they are ready to go, but probably they cannot indicate the particular food source that attracts them. Being limited to the here and now, primates cannot designate a time and place for meeting again after they have separated.

Apart from expressing his own intentions, a monkey or ape can *influence* the intentions and moods of his fellows. A baboon mother invites her infant to jump on her back by lowering her rump, and a male chimpanzee can calm a scared female by patting her extended hand. As with humans, the meaning of a signal must often be specified by the context. A threat, by itself, simply says "Stop this or I'm going to attack." But "this" can mean being too close or too far from the actor, or grooming a third animal. If the context allows more than one interpretation, the other primate will learn from repetitions what exactly he should not do.

Social Structure

The overwhelming majority of investigated primates live in more or less closed groups that are tied to a particular piece of land and that mutually avoid each other. These groups number anywhere from 2 to 700 individuals, but their central tendency is between 10 and about 80 animals. In most species, social interactions between members of different groups are very rare; females, especially, hardly ever meet a stranger. In contrast, interactions within the group are usually unlimited, and spatially distinct closed subgroups are uncommon (hamadryas baboons are exceptional in this respect). Thus, the typical primate population is organized on one major level, the group, while supergroups and subgroups are less clearly differentiated. The only pervading type of subgroup consists of a mother, her infant, and quite often its older siblings.

Each group lives in a home range whose central area is rarely entered by neighboring groups. Seasonal migrations are small in scale and restricted to the group's own home range. The group not only avoids unfamiliar neighbors, but also hesitates to enter unfamiliar areas outside its usual range. Thus social and spatial mobility is typically limited. Nevertheless, some adult males occasionally leave their group to live solitarily or with another group for days or months. In one forest-living population of anubis baboons in Uganda, this is in fact so common that groups can be defined only by their stable female membership. Thus, the degree to which a group is closed varies considerably.

In some forest-living primate species, the relationship between groups is specialized into so-called territorial behavior. Instead of avoiding each other, the territorial groups actively converge at a point near their common border for a session of basically hostile displays. The South American howler monkeys only roar at each other on such occasions; among Asiatic gibbons, calling and chasing among the males are combined; grey langurs in Ceylon actually fight. The groups of these species typically live in rather small home ranges, called territories. Since animals generally are more aggressive near the center of their range and more inclined to turn around and run back near its border, the difference between territorial (aggressive) and non-territorial (avoidance) behavior may often be based on mere modification, in which the size of the home range shapes the nature of intergroup relationships. Nevertheless, territorial displays are so highly evolved in some species that their territoriality appears to be a well-established phylogenetic adaptation.

The notable exception to the one-level, relatively closed spatial group is the chimpanzee. It is as yet uncertain whether chimpanzees have any stable groups at all, but investigators agree that chimp parties constantly change in size and composition. When such parties meet, they show neither avoidance nor aggression, but exchange friendly greetings. Chimpanzees (and possibly the other great apes as well) seem to recognize a far wider nexus of bonds and relationships than monkeys, which Reynolds (1966), in a remarkable article on this subject, called a "sense of community." Chimpanzees lack what to me is one of the most striking aspects of social organization in baboons: the powerful discrimination between members of the group and strangers, between "ins" and "outs." Of two conspecifics which present nearly the same stimuli, one is accepted with friendly lip-smacking whereas the other is violently chased away. It is as if a baboon dealing with his conspecifics distinguished between two entirely different classes of beings whereas a chimpanzee, in contrast, seems to recognize only one. Future research will have to show whether this picture is entirely correct.

In their composition, primate groups fall into three major categories. Most common, it seems, is the multi-male group in which many males and females live without stable heterosexual bonds. This type is common to all macaques, many langurs and baboons, and most South American monkeys.

The second category is the one-male group, which consists of one and only one adult male and several females. This is the group type of three open-country species, the savanna-living patas monkey, the hamadryas baboon, and the gelada baboon of the Ethiopian mountainous grassland. In the two latter species, many one-male groups live together in a large troop. In addition, one-male groups are found among some langurs and the forest-living guenons (genus Cercopithecus).

The third category is known only among the small apes (gibbons and siamangs) and the tiny South American titi monkeys (Callicebus moloch). In these species, the group consists of a single pair of adults. "Monogamy" thus is an exception in primates. Groups of all types, of course, include juveniles and infants. Solitary males and groups consisting only of adult males are found living apart from the heterosexual groups in several species.

Although a primate has access to all members of his group, he usually shows marked preferences for some members while hardly ever interacting with others. Such subgroup formation is partly based on unexplained individual affinities and partly on kinship and age-group preferences.

Nonhuman primates recognize only matrilineal kinship. The subgroup of a mother and her children shapes the latters' social relationships in three species investigated in this respect and presumably in many others as well. Chimpanzees preserve strong bonds with their mother and their siblings well into adulthood. The social rank of juvenile rhesus monkeys among their age group precisely corresponds to and depends on the rank of their mothers. A subadult rhesus male is less likely to leave his group if his mother still lives in it. New food habits introduced to groups of Japanese macaques most readily spread among siblings, from mother to child, and from child to mother.

Primate age groups appear to be socially most effective among males, as seen in play groups, which tend to unite males of the same age. The age class of subadult males tends to become socially isolated. Subadult hamadryas males, for example, spend only one-fifth of their time at resting places in social interactions, whereas juveniles and adults are socially active during one-half of that time. In many species the male subadults live on the periphery of the group and may even leave it.

Because of the general lack of detailed group histories, very little is known about primate "marriage systems." Research on the inhibition of mother-son matings in rhesus monkeys indicates that incest prohibitions in human societies may have parallels in nonhuman primates. Rhesus males rarely mate with their mothers because the son retains an inferior, infant-like position toward his mother and because of a specific inhibition that is independent of rank and thus restricts males even if they are dominant over their mothers. Both factors must cooperate to suppress mother-son matings (Sade 1968). In hamadryas baboons, father-daughter mating apparently is reduced by the fact that the daughters are carried off by young adult males before they are of sexual interest to the father.

As in many vertebrates, the behavioral differentiation of the sexes involves more than mating behavior. Primate males are generally more aggressive than females; their superior size raises their dominance status above that of most females of the group. Only males leave the group, become solitary, or visit other groups. Hamadryas baboons show similar sex differences already as infants: infant males leave their mothers more frequently than infant females. Young male monkeys of many species have a stronger tendency to associate with their own sex than have females. A study of Japanese monkeys suggests a further, rather subtle difference: When confronted with an unexpected novel object on their track, juvenile and adult females give obvious startled and avoidance responses. Adult males, in contrast, veer off to an inconspicuous angle when still at a great distance from the object and avoid making any conspicuous response.

It is hardly necessary to emphasize that at least when compared with humans, the open-country primate knows little of privacy. Group members may avoid the gaze of a particular individual, but as a rule, they are constantly stimulated by others who surround them. In such a society, the inhibition of uncontrolled social action can be critically important. On the other hand, a primate is rarely confronted by strangers. He lives in a society whose members and their probable responses in most situations he knows. Many mammal groups assemble, change composition, and dissolve again with the transitions between reproductive and nonreproductive seasons; primates maintain the same social organization with the same group members throughout the year. Thus, the uninterrupted, individualized group can become an ever-present tool of survival.

Such constancy, however, has its price. In the primate group, all social processes are continually going on together. There are no seasons set aside for dominance struggles among males, for mating, child-raising, or migration, each one accommodated by a suitable social arrangement and particular moods. A primate group must contain all these activities and their interferences within a single social structure. In turning from one to another context at a rapid rate, the individual primate constantly adapts to the equally versatile activities of the group members around him. Such a society requires two qualities in its members: a highly developed capacity for releasing or suppressing their own motivations according to what the situation permits and forbids; and an ability to evaluate complex social situations, that is, to respond not to specific social stimuli but to a social field. In the absence of privacy, which would mitigate both demands, primates heavily depend on these abilities.

Ecological Techniques

Social behavior in primates frequently reminds us of human parallels, but an observer of primate subsistence techniques will not risk drawing an analogy with even the simplest human economy. Subhuman primates have in common distinctly nonhuman ways of finding and using food and shelter. Indeed, the lower primates seem less "human" in these respects than many rodents and carnivores.

Most primates do not shape their environment in any adaptive way. The sleeping nests of the large apes are poor, roofless constructions built for only one night. Monkeys use their sleeping lairs exactly as they find them, without making any artificial alterations. No higher primates are known to gather or store food, and therefore they have no stable home base which could accommodate a sick or wounded group member for a day or two. The cursoriness of what actually is the most developed storing behavior in primates apparently impressed Pliny when he wrote: "The sphingion [probably the gelada baboon] and the satyr stow away food in the pouches of their cheeks, after which they take it out and eat it piece by piece; and thus they do for a day or an hour what the ant usually does for the whole year." Only chimpanzees carry food for as far as a mile. The hand-to-mouth economy of primates forces every one of them to seek food and water every day. When natural resources fail, they starve. In their rare hunting endeavors, baboons show no sign of cooperation and compete for the kill. One must turn to chimpanzees to detect signs of concerted action in catching the prey and sharing the meat. A party of adult male chimpanzees was observed by Jane van Lawick-Goodall to hunt a Red Colobus monkey. A chimp male occupied the foot of every tree that was connected at its crown with the one in which the Colobus sat, and another male climbed the monkey's tree to catch it.

Economically, primates have remained the unspecialized mammals that their anatomy declares them. This puzzling fact can perhaps be explained as follows: Primates lack the precise genetic programs for the complex behavioral sequences that are so remarkable in social insects and, to a lesser degree, occur also in rodents In an animal with a monkey brain this lack of programs merely results in a lack of complex sequences of behavior. It takes a brain that can *invent* a technique to benefit from the flexibility permitted by the absence of detailed genetic instruction. The human brain is of this kind, and tool-using in chimpanzees (such as the use of leaves as toilet paper) indicates an unmistakable though limited capacity for technical invention. Monkeys, however, have no advanced technical skills at all. Their evolutionary position in this respect is the technological no-man's-land between the genetically programmed techniques of a squirrel burying nuts at the foot of a tree, and the flexible insight of a man storing apples at the right temperature.

Thus, primates seem to have only one unusual asset in coping with their environments: A type of society which, through constant association of young and old and through a long life duration, exploits their large brains to produce adults of great experience. One may, therefore, expect to find some specific primate adaptations in the way primates do things as groups.

SUMMARY

1. A hamadryas troop is divided into bands, and bands are composed of one-male groups. The daily course of activities shows these three levels of social organization in their respective functions. The units of the same level vary in the degree of their mutual tolerance, and important parts of social behavior regulate the balance between unit identity and unit compatibility.

2. Primates can communicate only about the here and now. The exact meaning of a signal must often be inferred from the context and from previous experience of similar situations.

3. Most primates are organized in relatively stable groups that are tied to a piece of land. Unfamiliar terrain is avoided, and neighboring groups are avoided or engaged in hostile displays. There are three basic social structures: groups with many males, one-male groups, and monogamous pairs. Males are more migratory than females. The lack of privacy and of seasonal defferentiation in social life require a high degree of behavioral integration in the individual.

4. Nonsocial ecological techniques are poorly developed in primates. Their specialization must be sought in the way they act as groups.

Chapter 3

ADAPTIVE FUNCTIONS
OF PRIMATE SOCIETIES

THE TERM "GROUP" IN PRIMATOLOGY

In defining the primate group we must accept the limitation of criteria which transform every anthropological term when it is applied to animals. All subjective phenomena of group life—like "identification" or "identity of aims"—are lost for the study of animal groups, because their members lack the symbols for their expression. Definitions of animal groups are restricted to so-called objective criteria because they are the only criteria available. Only the student of human societies is in the enviable position of using and comparing both kinds of characters, what people do and what they feel.

Most students of primate behavior would agree that there are two proven parameters of describing the grouping tendencies of a population: the distribution of its individuals in space, and the frequencies and types of directed communicative acts among them. The use of these parameters may be illustrated with the one-male groups within a troop of hamadryas baboons. Repeated estimates of distance show that the mean distance between a female and her male in the resting troop is about 2 feet, whereas the average distance between two troop members chosen at random is about 10 yards. The members of a one-male group obviously follow each other closely and thus remain together within the crowd of a hundred or more troop members. Even so, a photograph of a troop rarely reveals the one-male groups as spatially separate units, since usually no empty space is left between them. The parameter of communication, however, shows that the one-male groups are distinct units even with regard to their nearest spatial neighbors: The adults of a resting one-male group socially interact with others in about 50 of 100 observation minutes. Of these 50 minutes only 3 contain interactions with troop members outside the actor's own one-male group; the other 47 minutes are spent in interactions *within* the group (Fig. 3.1). Thus, a primate group can be defined as a *number of animals which remain together in or separate from a larger unit and mostly interact with each other.* Usually, one is also interested in a group's stability over time, that is, in the persistence of the two parameters.

This definition merely permits us to identify and describe primate groups. Description must be followed by investigations on causation and function, two aspects of living things which biologists fervently try not to confuse. This chapter deals exclusively with the function of groups: It asks what group life does for the species, not why and how it comes about.

GROUP SIZE AND THE RESOURCE UNIT

Among the few well-quantified facts on primate social life is the size of their groups, and we may begin by asking how the number of individuals in the group may affect its success in daily life.

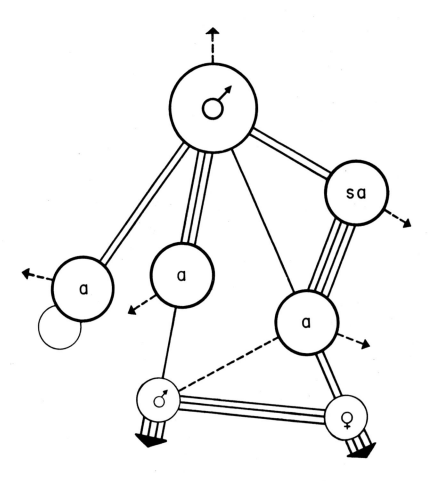

Fig. 3.1. Sociogram of a hamadryas one-male group. The number of lines connecting the individuals represents the frequency of social interactions among them. The broken arrows indicate that the male (♂) and his adult (a) and subadult (sa) females interacted with outsiders in less than 3% of the observation minutes. In contrast, the two juveniles (bottom of graph) scored 40% in interactions with other groups.

In order to survive, one must exploit resources and avoid injury. For a monkey or ape, this includes drinking and eating, on one hand, and avoiding predators and using safe roosting lairs, on the other. Among primates, every group member has to feed, drink, flee, and climb a tree for himself. There is no sharing or passing of gathered food or prepared shelter. Mutual assistance is negligible or absent. Only infants are carried and nursed. In such an economy, the group seems materially useless.

However, before the primates can react to resources and dangers, these must be located, and this is group life's first important function. Primates do not exchange resources, but they exchange information about them. For example, the discovery of a small waterhole by one baboon is immediately communicated to his neighbors by the unique posture of the drinking animal: head low, rump and tail in the air. The agitated hand movements of an animal digging up preferred food also attract his fellows. But discoveries of one group member benefit the others not only there and then. During a time of drought, an old male may lead the way to a distant pond which he remembers from a long-past visit in a similar situation. It may not be accidental that primate males, who are generally more active than females in leading the group, are also more likely to travel far and alone, especially when they are young.

How does group size affect the success of sharing information? It seems at first that the larger the group the better, since the amount of shared information would be roughly proportional to the number of members. But actual group sizes indicate that there must be limiting factors. The most important among them is probably the distribution of resources in the habitat. In discussing this, we shall for the moment disregard all factors except food.

When a monkey finds a small immobile food item, such as a nut, he will feed only himself; there will be no use broadcasting the discovery. If such nuts were regularly distributed over the entire range, no individual would ever learn anything about the location of nuts that the others would not know already. There would be no information to share. This may be one reason for the relatively small group size of primates in arboreal forest species, where food seems more evenly scattered than in savannas. If, however, the nuts occur only in certain spots and in quantitites that could feed more than one monkey, then the group will benefit from his sharing such information.

My first thesis is, therefore, that foraging in groups is adaptive only if food appears in clusters or units that can feed more than one animal. We approach the question of group size by asking *how many* individuals can simultaneously feed on such a food unit. There is no point in attracting twenty group members to a little bush that can feed only three.

There are three ways in which primates appear to solve this problem. First, most species use no calls when they find food. Thus, primates who find food attract the attention of only those group members who can see them at that moment. The message spreads slowly and reaches the entire group only if the food is abundant enough to last for quite a while. The only exception known to me are the chimpanzees of the Budongo forest in Uganda. The tremendous calling and drumming raised in this population when a large stand of fruit-bearing trees is discovered seem to attract other parties.

The second mechanism that limits clustering is that feeding primates tolerate others only beyond a certain distance, usually a few feet. These distances are generally recognized by the inferior group members, who will, therefore, look for another food source when the first is fully occupied. This technique is satisfactory in a relatively rich habitat where food units are not too far apart from each other. It is a poor solution in a barren area where food sources are widely separated. The inferior animals would be left outside and hungry at each food source the group would pass.

In such conditions, a third solution is appropriate. It consists in adapting the size of the foraging group to the size of the food units by means of social organization. My thesis is that under harsh feeding conditions—and this is where we must expect the most precise adaptive fit—group size will closely approach the maximum number of animals that can simultaneously feed on the most crucial type of food unit.

The actual group sizes, however, vary widely even within the same species and in the same type of habitat, and each group maintains its size through all its activities for many months. These data are not very helpful. There are, however, a few species that *change* the size of their group according to their type activity, such as the chimpanzees, the geladas and the hamadryas. Here, we may find some clues.

In hamadryas baboons, the correlation between group size and resource unit is relatively clear. In their Danakil habitat, the flowers and beans of small acacia trees are the main food source. One regularly observes that an isolated tree is picked by a single one-male group, about five animals. This number allows the baboons to keep their usual feeding distance of several yards, at which low-ranking animals are not impeded by their dominant neighbors. The one-male group thus appears as the "single-tree-foraging unit." The groves of ten or more large acacias on the larger river beds are usually occupied by one band at a time. In the dry season, waterholes become the critical resource unit. These river ponds are then situated miles apart, but most of them carry enough water for a hundred or more baboons. Though hamadryas troops rarely assemble on their daily route, and never assemble to drink during the rains, they do assemble at waterholes in dry months.

The second important observed function of troops is their optimal use of sleeping-cliffs. Hamadryas cliffs, a large resource unit, are scarce in some areas, where, unless the population makes full use of them, roosting-space becomes a population-limiting factor. In such areas, hamadryas form troops of up to 700 animals, aggregations that are much larger than suitable for any foraging unit in the area. Comparisons of troop size in areas with many and few cliffs bears this out: In the Awash Valley, where food is scarce but canyon cliffs are almost continuous, troops are one-fifth to one-tenth as large as the troops in the nearly cliffless country around Dire

Dawa, where the food situation is improved by agriculture. Although these hypotheses await quantitative testing, we may speculate that there would be no troops among the hamadryas if their roosting-lairs were small and widely separated, each accomodating but a few animals. Anubis baboons, who have sufficient roosting-trees in their typical habitats, do not form troops but make up groups the size of a hamadryas band. When drought forces anubis baboons to use the same waterhole, the groups approach each other with caution. The society of hamadryas baboons thus includes social units of three sizes, each seemingly fitted for a particular type of resource unit.

Patas monkeys (Figs. 3.2 and 3.3) offer another example of the relationship between group size and the size of the roosting facility. A population studied in Uganda lives in a savanna with no cliffs or high trees. The patas monkeys must roost on trees which apparently are too small even for their one-male groups. Instead of merging for the night like hamadryas, the patas one-male group splits up to sleep singly or in pairs over a wide area of the savanna (Fig. 3.4). If patas one-male groups meet at waterholes, they threaten and chase each other. Their social behavior is programmed for dispersal, while that of hamadryas baboons is designed for both dispersal and fusion. The gelada baboons of the treeless Ethiopian highlands (Figs. 3.5, 3.6, 3.7) are of flexible group size. In this species, the society is organized on two levels, the large troop and its component one-male groups and all-male groups. In areas and seasons of good food supply, the geladas tend to form troops. When food is scarce, the troops split up into the smaller units which live and forage independently.

Chimpanzees are the third species that rapidly alters group size by fusion and fission. According to Reynolds, this system "allows dispersion and aggregation to exploit seasonal variations in availability and distribution of fruit," which would be compatible with our thesis. Chimps are, nevertheless, a special case in that they are the only known forest pri-

Fig. 3.2. Adult male patas monkey (Erythrocebus patas). Close relatives of the vervets (Fig. 3.3), the patas have specialized for terrestrial life by increased size, increased sexual dimorphism, and great speed in running. (Delta Primate Research Center.)

Fig. 3.3. Adult male vervet (Cercopithecus aethiops). The species is typically found in gallery forests.

mates that form a fusion-fission society. It is difficult to see why forest-living monkeys are not equally flexible but stick to one-level societies. The answer may lie in the causal instead of the functional dimension. As I shall show for the hamadryas, multi-level societies are difficult for monkeys to realize. The presence of smaller units in a larger group is a

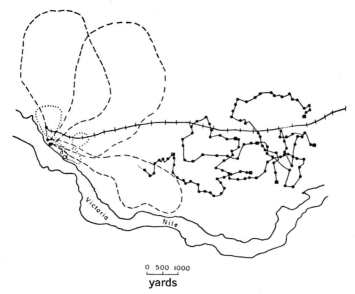

Fig. 3.4. A comparison of daily travel routes of baboon, vervet, and patas groups at Chobi, Murchison. Whereas baboons ($\bullet$ --- $\blacksquare$) and vervets ($\bullet$ $\cdots$ $\blacksquare$) return to the gallery forest every evening, the patas ($\bullet$ —— $\bullet$) remain in the savanna even for the night. (Reproduced from Hall, 1965.)

Fig. 3.5. Habitat of the gelada baboon (Theropithecus gelada) in the Semyen mountains of Ethiopia, at 12,500 feet above sea level. Open alpine meadows scattered with lobelias (foreground) are the foraging grounds; the nights are spent in the cliffs of the escarpment.

Fig. 3.6. A gelada troop traveling from one feeding ground to another along the edge of the escarpment (background). The troop is flanked by a protective party of adult males on the inland side.

Fig. 3.7. Heavily mantled adult gelada male walking at the periphery of a troop. Females (left) have no mantles and weigh only half as much as the males.

constant source of conflict, since units and subunits are stable, rigidly closed, and potentially hostile toward each other. Fissions and fusions are possible only along strictly determined seams. Among the apes, with their disposition for tolerance and truly open groups, foraging units change composition freely. This difference between monkeys and apes may perhaps explain why fusion-fission societies occur only in those monkey species where a harsh environment enforces them by unusually strong selective pressures.

In summary, deduction and examples indicate that group size is favorably adjusted to the size and dispersal of resource units. If the same habitat offers needed resources in units of varying size, it seems that its inhabitants would fare best with a flexible organization. There are several possible explanations of the fact that most primate species live in one-level groups of constant size. For example, resources may be evenly dispersed; or only one type of resource unit may be scarce enough to require an adaptation of group size; or the phylogenetic heritage of the species may not include the behavioral mechanisms that permit a multi-level society, in which case group size would approach a compromise between the requirements of several resources.

THE GROUP AND ITS PREDATORS

Predators are much more important in the ecology of nonhuman primates than in the life of man. While man belongs to the hunters, the smaller, unarmed monkeys are among the hunted. Terrestrial primates like the baboons cannot afford a division of labor that would separate the sexes for different foraging tasks. Without the presence of their large, canine-toothed males, the females probably could not forage without losses to predators.

The problem of predation is most critical for terrestrial, open-country primates. We know of two major techniques they have evolved for dealing with predators: the baboon way and the patas way. Among the usually rather noisy baboons, the emphasis is on discovering the predator, rather than avoiding discovery by it. Once contact with the predator is made, baboons may attack and mob it rather than fleeing indiscriminately and under every circumstance.

Concerted, defensive actions by a baboon group against jackals, dogs, or leopards have been observed repeatedly. Typically the large adult males interpose themselves between the predator and the rest of the group, barking at and displaying their impressive teeth by opening their jaws widely. In the Amboseli National Park, the Altmanns observed a leopard as it jumped from the undergrowth into a group of yellow baboons at the edge of a waterhole: "The baboons sprang away, then turned on the leopard, barking loudly as several members of the group ran at the leopard. At one moment, the dominant male was closest to the leopard. Faced with this mass attack, the leopard turned and ran." An adult male, a subadult male, and a juvenile male were wounded in this encounter, but all recovered. In another instance a chacma baboon group was threatened by dogs. The large adult males immediately interposed themselves between the troop and the attackers: "It was not uncommon for a single dominant male to maim or kill three or four large dogs before retreating in the direction taken by the troop" (Stoltz and Saayman, quoted by Altmann and Altmann 1970).

What, theoretically, is the optimal group size for these defense tasks? Unlike food, water, or roosting sites, the predator is mobile and can become a danger for many group members within a few minutes. The discovered predator merely shifts its place, whereas a resource unit, once discovered, is exploited and thus becomes irrelevant to other group members. This explains why the approach of a predator, in contrast to the presence of food, is signalled at the maximum distance the baboon's voice can carry. For the task of discovery, the ideal "anti-predator unit" is very large, since every pair of eyes helps and the more baboons reached by the discoverer's warning bark, the better. Often a discovered predator immediately leaves the informed group and tries its luck somewhere else. Its chances will be better if its prey is found in many small groups that are out of communication range rather than in one large group.

The task of mobbing and discouraging the predator also seems to favor the large groups of 50 or more generally formed in most baboon species. However, the reader will without difficulty find a score of variables that a detailed model would also have to include, such as the number and technique of cooperating predators, their visibility, the advantage of temporary large groups that can feed the existing predators with diseased or weak members alone, the negative effects of needless alarm over great distances, and so forth. Within the size range of baboon groups under savanna conditions, however, we may assume that the larger group is a safer group. Predator pressure thus will tend to increase the group sizes elected for resource conditions. The small foraging units of hamadryas baboons support our impression from the field that predator pressure in the semi-desert is less important than the food problem. But we nevertheless observe that the one-male groups forage together in larger, safer units wherever food conditions permit.

This pattern differs from that of the patas monkeys of Uganda studied by Hall (1965). The small, isolated one-male groups of these tawny, long-limbed monkeys are silent and furtive. They do not face an approaching danger as a group. Instead, the alerted male climbs into the upper branches of a tree from which he scans the area. From this exposed vantage point, his size and the white color of his thighs make him rather conspicuous (Fig. 4.5). His further behavior in the presence of a human observer suggests that his role is not only to watch for danger, but also to divert attention from the group. He noisily and conspicuously bounces on the branches, making them shake vigorously. He may also descend from the tree, produce a "whoo-wherr" growl, gallop very close past the observer, or run across the savanna far away from the intruder and the group. As the male engages in such distracting displays, the females and juveniles silently remain in their places, often lying flat on their bellies in the grass (Fig. 3.8). Their last resort in avoiding predators, however, is their tremendous running speed. In this respect, the patas monkeys clearly beat the baboons and possibly all other primates.

The patas technique of avoiding predators is to see but not be seen. While the first is best accomplished by a large group, the second function requires silence, a high degree of dispersal, and small groups. That the patas follow the second course is probably because their resource conditions favor small groups. The loss in the number of eyes is compensated by the degree of watchfulness of the male, who seems to be more alert than any single baboon male. Both in captivity and in the wild, the patas male usually stays at the periphery of the group, sometimes at a considerable distance, and often facing away from it.

Among the ground-living primates, the patas monkeys' well-defined roles of attention-diverting males and hiding females is unique. A somewhat similar role distribution is found only in some arboreal species of guenons (Cercopithecus). For example, when a one-male group of Cercopithecus nictitans is disturbed, its male emits loud calls and approaches the disturbance, while his group withdraws from him and the danger. Gabonese hunters provoke the approach of the group male by shaking branches and thus improve their chances of shooting him. Like the patas male, the nictitans male diverts the danger rather than leading the group.

The baboon way and the patas way of dealing with predators are relative, not rigid specializations. One day in the Awash Park in Ethopia, we observed a large group of anubis baboons running in a dense pack to cross an open plain. The baboons are not fast or persistent runners, so after only about 400 yards, an adult female dropped out in the middle of the plain, apparently neither wounded nor sick. Unable to flee at our approach, she chose the alternative technique of

Fig. 3.8. The patas monkeys' tendency to hide is manifest even in social contexts. Here, a subadult male patas hides behind a tree while watching a fight between other group members. (Delta Primate Research Center.)

crouching quietly in the concealing grass like a patas. On the other hand, adult patas males in West Africa have been observed actively defending their groups. In one case, a patas male chased a jackal which carried an infant patas in its fangs. The jackal soon dropped its prey, and the infant, apparently unharmed, was retrieved by a patas female.

These examples of foraging, roosting and predator-avoidance suggest that large or small groups can be ecologically advantageous in one functional context but maladaptive in another. When a population flexibly shifts between different group sizes, we can expect these group sizes to relate to several important factors in the environment. If only one type of group is maintained, its size may fit a single most important condition, or it may be a compromise.

Primatologists used to formulate hypotheses relating an average group size of species and the overall character of their habitat, such as "forest" or "savanna." Unless the various and perhaps conflicting requirements of these habitate are identified, such hypotheses are questionable. Most of them have not been supported by recent data. For example, ground-living species generally form larger groups than species living mainly in the trees, but there are exceptions, such as the extremely terrestrial patas monkeys with their small groups, or the forest-living Cercopithecus l'hoesti, which is more terrestrial but lives in smaller groups than other Cercopithecus species in the forest. Several arboreal monkeys tend to form mixed groups, consisting of two or three groups of different species. Apparently, some ecological factors favor large groups even in the forest. The tiny arboreal talapoin monkey and the large terrestrial drills are forest primates that have both been observed in groups numbering more than a hundred.

COORDINATION OF INDIVIDUAL ACTIVITIES IN THE GROUP

A major ecological value of the primate group is communication about resources—regardless whether the information concerns the location of a stand of grass ears that will be gone a minute later, or the position of a cliff that will remain valid information for a long time. Though the information is carried and passed on, the resources themselves are not. With the exception of defense against predators, each animal has to take care of his own needs, but it must do so within the group, and this is not quite so easy as it first seems.

Imagine a group living in an ideal forest habitat where food, water, and safe places to sleep occur in small portions that are densely and regularly distributed over the area. Under such conditions, each animal finds what it requires within a few steps. It does not have to leave the group to satisfy any need. It can eat, drink, and sleep at any time, quite independently of what others are doing.

Take now a similar group in a different habitat, where resources are clustered and the clusters are far apart—so far, in fact, that communication between neighboring resource sites is impossible. When a group is feeding at one site, a thirsty member cannot go to the nearest water without losing the group. In this situation, group life demands that all its members do the same thing at the same time, that they adhere to the same schedule. There is a possible danger here: If an animal does not drink when its group seeks out the waterhole because it is not thirsty yet, it will suffer from thirst before the group reaches water again, unless it separates from the group.

Group life under such conditions requires a secondary adaptation that makes the primary adaptation of group life possible: It depends on a behavioral mechanism that induces the individual to do as his fellows do. Such a mechanism, called *social facilitation* by ethologists, is widespread among all social species. It makes young chicks eat more when they peck in company, birds fly off in flocks, humans get angry or yawn when they see others do so. Social facilitation synchronizes activities which, as such, could very well be carried out individually and at different times. It can be observed or expected whenever it is advantageous that every group member take up the activity of the majority.

Social facilitation should not be confounded with the much more difficult behavior of *imitation*. When a human yawns in response to another's yawn, he does not need to concentrate on the performance of his partner in order to know how to yawn. He has the genetic program for yawning, and he could yawn even if nobody ever showed him how it is done. That in social facilitation the releasing stimulus coincides with the released response is, as it were, accidental and does not improve the quality of the response. (The reader may by now begin to yawn himself with nothing to act on but the written word.) In social facilitation nothing is learned, and the response can be unconscious. In contrast, imitation is the copying of a novel or otherwise improbable act for which the imitator has at best small fragments of a genetic program. Learning a difficult dance is an example. Imitation requires close and most probably conscious attention. Most birds and mammals show social facilitation, but limitation is found only among humans, the great apes, and, to a very limited degree, in monkeys.

One would expect the highest degree of synchronization in vital activities such as fleeing; and indeed, a startled response of a baboon is taken up by his neighbors within a fraction of a second. Feeding, drinking, and beginning to move spread at a more leisurely rate. Warning calls and mobbing behavior are also "contagious," but juveniles and females do not always participate. A monkey scanning the more distant surroundings from a tree will usually provoke no similar behavior unless he signals a discovery. Thus the degree of coordination in the group seems related to two factors: how critical it is that every individual perform the activity and the time available for the response to take place.

In an extreme case, a single group member can take over a "task" for the entire group. Leading the group and scanning the distant surroundings are social functions that make group life most advantageous. In terms of the group's time budget, it would be inefficient if most members constantly sat on barren termite hills and looked for leopards. One or two scanners plus the limited alertness of their foraging fellows must suffice. In these cases, group members differentiate their action patterns; they assume "roles."

It is interesting to ask whether there is a counterpart to social facilitation in such one-for-all tasks, a *social inhibition* that would reduce an animal's tendency to perform an activity when it sees another group member already attending to it. With respect to temporary roles like sentry activities, the question has not yet received specific attention by field students, but it is known that permanent roles in the group are subject to such mechanisms. When a group has a leader, another male may not lead the group or scan the surroundings until the day that that leader dies, but when this happens, he may take over the leader's role within hours. One must assume that the previous leader's activities inhibited leading behavior in other members of the group. Such discouraging effects are quite obvious in some social roles. Macaque mothers or hamadryas harem males go so far as aggressively to prevent others from sharing their roles toward their infants and mates.

Thus, social order provides a complete scale of coordinating mechanisms, ranging from strong encouragement to violent discouragement of "doing likewise," depending, it seems, on the number of animals that a task requires.

DOMINANCE

The extreme form of social inhibition is known as *dominance*. Its ecological function is to clarify the situation when the same action cannot be carried out by more than one group member. When a resource unit—a fruit-bearing twig or a sleeping-ledge—is so small that only one animal can use it, the more dominant animal will take it.

The term "dominance" is widely used to describe a particular type of order in organized groups. Its most general criterion is the fact that an animal consistently and without resistance abandons his place when approached by a more dominant group member, a sequence called "supplanting."

In primates, older and stronger individuals supplant the weaker ones, and males are generally dominant over females of the same age. Each group member has to learn his rank of dominance, at least within his own sex and age class. The ultimate means of clarifying positions is fighting.

The consistent ability of some animals to supplant others has several effects on the individual's ecological prospects within the group. The dominant primate can displace his inferiors from the best feeding sites and the safest sleeping places. Thus, baboons sometimes supplant low-ranking group members from grass plants that have already been dug up, by a subtle approach-avoidance sequence without any threatening gestures. This advantage is usually of little importance, since primate vegetable food occurs mostly in small bits scattered over an area that can accommodate all members of the foraging party. With large items of food, however, dominance becomes decisive. The young antelopes that baboons sometimes kill are almost exclusively eaten by the adult males, and fighting over such prey is frequent. The inability of baboons to share food is a behavioral characteristic that probably prevents them from shifting to hunting as a way of life. In contrast, chimpanzees, who also kill and eat small mammals on occasion, do beg each other for parts of the prey and are sometimes successful in obtaining them.

Even with a vegetable diet, the effects of dominance become critical when the total available amount of food falls short of the group's needs. There is a wealth of data, mainly from studies of animals in captivity, demonstrating that under such conditions the dominant animals will take the food while the inferior ones will suffer from the shortage. Dominance will then probably favor the adults and sacrifice juveniles and infants. This seems adaptive, since the experienced and reproductively active adult is more valuable to the group than an easily replaceable youngster. But a food shortage is also critical for the females, placed at a disadvantage by size and dominance factors, and far less replaceable than the youngsters. At present, we can see two possibilities for counter-designs that would protect the females. One speculation, put forward by Crook (1970), suggests that the one-male group of the open-country species is a way of freeing the females as much as possible from food competition with males. Among patas and geladas, the females live with a single male, who is essential for breeding and protection. The supernumerary males of these species form purely male groups that forage outside the home range of the one-male group.

Another speculation can be based on the fact that primate males are heavier than females in most species living in meager habitats. Among hamadryas baboons, females and juveniles often save themselves from a male attack by rushing onto a limb so thin that the heavy male will not dare to follow. Smaller animals are obiously carried by lighter limbs than heavy ones. Even after a 40-pound hamadryas male has exploited an acacia tree for flowers or seeds, a 20-pound female will still find food on it. A future field study may investigate the hypothesis that, in a species that obtains much of its food on the ground, arboreal diet becomes the critical food source of females and juveniles because it reverses the advantage size has on the ground.

The great sexual dimorphism of terrestrial primates is usually interpreted as an adaptation to defense against predators. This is probably the major selective factor in a species like the geladas, which have never been seen to feed on trees. The argument is less convincing for the terrestrial and equally dimorphic drills and mandrills; in their dense forest habitats, they are never forced to face a predator on the ground. In these species and in the savanna baboons, the weight-in-trees factor may have contributed to dimorphism.

The dominance order has certain ecological advantages, as in substituting a smoothly functioning order of priority for endless quarrels about resources. Another function of dominance is the dispersal of the group members in space. The less dominant members avoid entering the area around a dominant animal. This tends to scatter the group, so that foraging individuals will look for food at distances that keep them from interfering with each other. At larger distances, however, dominance seems to reverse its spacing effects. The British ethologist M. R. A. Chance (1970) has pointed out that the non-dominant group members tend to anticipate the movements of those higher in rank. In order to do so, they frequently look up and check on the whereabouts of the most dominant members, even following them at a distance to keep them in sight. This focused attention, together with the protectiveness of the dominant animal, make him an attractive figure at distances of 5 or more yards. Because of the great attention paid him by the rest of the group, the highest in rank becomes a potential leader whose actions are likely to influence others. This hypothesis, again, awaits experimental testing.

The most intricate effects of dominance are not ecological but purely social. They are causal rather than functional aspects of primate societies, and thus fall outside the frame of this chapter. Ecologically, the position of the dominant animal is of little significance unless it is associated with leading or protective functions. Consequently, thinking in terms of dominance is currently being replaced by thinking in terms of roles. A role, however, should not be described merely as what an animal does, but rather as an individual, group-oriented function. Thus, Hall likens the patas male to a "watchdog" of his group. This extension of the laboratory-bred term of dominance opens the way for fresh research on the mechanisms of role distribution and role differentiation, on mechanisms more subtle than open competition, and on the possibility that group members may even compete for roles that benefit the group, not just contribute to their own survival.

LEADING THE GROUP

The ecological function of leadership is an obvious example of a social role in primates, although it has not yet attracted topical research projects. Ecologically, the importance of leadership increases with the distance between resource units. In the tiny range of a titi monkey group (Callicebus moloch), where food trees are evenly scattered, leadership is hardly important because one layout of the daily foraging trip is about as profitable as another. The group traverses its whole range, which is less than 110 yards in diameter, several times a day anyway.

The situation differs greatly for some population of chimpanzees and hamadryas baboons. The latter cover an average of 8 miles a day, a distance greater than that of any other primate species investigated. The roosting-cliffs of hamadryas are miles apart, as is the case for their watering places in the dry season. The home range of a typical troop has dense stands of acacia bush, the staple food source, though these are separated by long stretches of meager, stony grassland that is completely dry for several months of the year. Under these conditions, the effort of foraging may exhaust the time and energy of the weaker troop members. The 6 to 12 miles of daily travel covered by our troops in the Danakil plain may be nearly at the upper limit of a small juvenile's capacity. Tolerably long routes that begin and end at a cliff, touch on water at least once, and lead through one or two satisfactory feeding areas are not too many.

Thus daily foraging of this kind calls for a function analogous to planning. This in turn requires that at least some troop members are informed about the whereabouts and present conditions of the resources. To obtain the ideal design of the route, the slightly differing information of all members would have to be pooled and evaluated. This, however, is impossible, since primates cannot communicate about distant food groves and waterholes and their recent condition. As already mentioned, all a primate can do to inform the departing troop of a profitable location is to indicate, by the direction of his glances and shifts, the direction which he intends to take, and possibly the strength of his motivation to go there.

An alternate procedure would be to follow one well-informed leader. Surprisingly, this appears to be a rare solution among primates. Among anubis and yellow baboons, for example, travel is initiated by many individuals, males and females in turn. Leadership by one adult male is more common in species with one-male groups. Among geladas and hamadryas baboons, the one-male group is commonly led by its male, but even here, females may temporarily take the lead. In the one-male groups of the strongly dimorphic patas monkeys, the females direct and initiate group travel more often than not while the male takes up his peripheral lookout positions.

Mountain gorillas provide the clearest example so far known of leadership concentrated in one individual. This single silverback male is usually in the lead position when the group travels rapidly. Before departure, the leader sometimes "employs a characteristic posture which apparently serves as a signal to the other members of the group, indicating his imminent departure. He faces in a certain direction and stands motionless for as long as ten seconds with front and hind legs spread farther from each other than usual" (Schaller 1963). On occasion, he emits a few short, forceful grunts, to which the group responds by moving in his direction.

As a rule, however, primate groups are led jointly by several adults. This means that the actual route is the result of compromise. Surprisingly, it is nearly impossible to discover signs of conflict between group members with different directional intentions. In the face of ecological necessity, primates have as yet never been observed to dispute the course of their action by even the mildest threat. Traveling group members can nevertheless tend in different directions and follow their designs quite inflexibly. I once observed such a struggle in a small hamadryas band that consisted of only two one-male groups, led by two males, whom I had named Circum and Pater, respectively. The band departed from its cliff shortly after 7 A.M. The younger of the two males, Circum, immediately "proposed" a northerly direction. My field notes contain the following account:

07.30   Circum utters a contact grunt and goes *northward* along the riverbed. Again the entire party follows for some 20 yards and then stops.

07.31   Circum again rises; he briefly looks back at Pater and then goes another 30 yards northward. No one follows. He stops, comes back until he is only 20 yards away from his closest female and sits down. All the while Pater has been watching him.

07.32   Circum rises and begins to move *west*, straight across the river bed. Only his youngest female follows him. After a few seconds both come halfway back and sit down.

07.33   Circum sets out again, this time in a *southwest* direction. Now, Pater rises, and the whole party follows Circum in the same marching order as above.

07.40   On the left bank, Circum again begins to point northward. The others follow him. After 110 yards they all sit down, then they climb an acacia and begin to eat the blossoms.

07.58   Pater climbs down and sits next to the trunk facing northward. Circum immediately climbs down to him; Circum stops a second near Pater while the two of them turn their faces toward each other. Then Circum continues further north. Pater allows females and young to pass and then follows at the rear. Twenty yards ahead, everyone stops.

08.07   Circum rises, looks back and proceeds. All follow. After a few yards he swerves to the right, leads the troop back to the right bank, and continues northward. At this, Pater slowly moves to the front of the column. Ten yards ahead, he overtakes Circum and as he reaches the front, turns westward back to the left bank. The females and young swing around him to the west like a rope. Circum, on the rope's end, continues northward a few paces but then follows the rest across the river. For the first time the marching order is reversed: Pater, Pater's only female, Circum's females, Circum. The young are scrambling about at the side of the column. Pater now leads the party westward and then to the southwest. (Note that this was the direction in which he ultimately followed Circum at 07.33.)

During most of the day, Circum tried to lead the band to the north, where the rest of the troop had gone a few days before. He was adamantly opposed by the older and more influential Pater, who insisted on a trip to the southwest. As a result, the trip assumed a peculiar zig zag pattern, but the two males never showed any signs of impatience or aggression, nor did the two groups separate. At 2 P.M., Circum finally abandoned his northward trend and preceded the band in the direction that seemed to correspond with the intentions of the older male.

Quite another question is how information on the current state of food sources is obtained. When howler monkeys arrive at a point of choice in an unfamiliar area, all the adult males simultaneously explore alternate parts of the arboreal paths. "When one of them finds a suitable route, he will give the deep clucking vocalizations; then the females and young begin to follow him slowly, and associated males cease their exploratory behavior and fall in line with the moving column of animals" (Carpenter 1964). Thus the less familiar parts of the howler monkey's route at least are not decided upon before departure; the alternatives are explored on the spot. In a howler territory, which is only about 650 yards in diameter, this is an acceptable solution. For hamadryas baboons, the distances to be covered by exploring males would obviously exceed the tolerable; the explorers, moving far beyond contact with the waiting troop, would have to return to make any clucking noises. Aside from the time factor, it is uncertain whether the dominance-oriented baboons would be capable of a howler-type decision process, which, in effect, seems to chose the most attractive pathway regardless of the dominance status of its discoverer.

Chimpanzees, however, seem to use the design of the exploring party efficiently. I have already quoted Vernon Reynolds, who reports that feeding chimps in the Budongo Forest of Uganda can attract other parts of the community over distances up to 2 miles. Parties of chimpanzees meeting at a tree with abundant food engage in a chorus of loud calls and resounding beats against the plank buttresses of trees. Reynolds suggests that their vigorous drumming communicates the food location to other parties in the area.

In the semi-desert, the extent of a daily trip exceeds the

carrying distance of even the loudest bark of an exploring party. In choosing the direction of their departure, hamadryas baboons have to rely on the information gathered on preceding trips. We do not know how and by whom the sites are explored and remembered, but we know that on most mornings different males of the troop strive in different directions. The decision is made by a long process in which most adult males of the troop participate while the rest seem unconcerned. I have already described how a hamadryas troop prepares for departure during its morning rest. The troop performs slow on-the-spot movements, changing its shape like an undecided ameba. Here and there, males move a few yards away from the troop and sit down, facing in a particular direction away from the center. Pseudopods are generally formed by the younger adult males and their groups. For some time, pseudopods protrude and withdraw again, until one of the older males in the center of the troop rises and struts toward one of the pseudopods. At this, the entire troop is alerted and begins to depart in the indicated direction. Detailed observation reveals that two male roles are involved in leading the hamadryas troop: the younger initiators who "propose" certain directions and the deciders who choose among the proposed directions. Accordingly, we can talk about the "I-D system"; whereas the troop as a whole pays little attention to the initiators, it immediately follows a decider. The I-D system is also found among the forest-living anubis baboons of Uganda. In this population, however, the deciders are adult females, the males being restricted to the role of initiators.

Interestingly, the I-D system is not a rigid trait of hamadryas organizations. In 1968, when we restudied a troop of hamadryas baboons which had shown the I-D process very clearly in 1961 and in 1964, we found almost no trace of the former routine. Instead of protruding and retracting pseudopods, the troop now departed without much ado from its resting place above the cliffs. Males seemed to disagree less on the direction of the troop's departure. We have no explantation for this change. There is certainly no need to postulate a new tradition. The food situation may have changed or, possibly, an influential leader could now lead the troop without much resistance from other males. Field studies have amply demonstrated that primate species show geographical variations in their organization. Our experience indicates that even the same troop can change its social behavior according to the set of qualities of its actual members. A group process can take a different form in a new generation without any underlying changes of gene pool or tradition, simply because each generation of group members is a new combination of old traits.

In the most complete ecological report on a primate species now available, Stuart and Jeanne Altmann indicate that the yellow baboons of Amboseli National Park may change the utilization of their home range according to what they have experienced in certain locations: "One [sleeping] grove . . . was deserted after two members of the group were killed therein by a leopard. Although this grove had been used as a sleeping grove on 13 of the 57 preceding evenings . . . , it was not used once in the subsequent 68 nights for which we have sleeping grove data."

Apart from selecting a nourishing and safe travel route, the group faces the task of maintaining contact even when visibility is poor. When a hamadryas band travels intermittently in a long scattered column, each family male tends to sit between shifts on an elevated spot from which he can see both the preceding and the following males. When a male moves on, the following unit leader usually comes up and sits in exactly the spot just vacated by his companion. This relay behavior seems to support group cohesion in dense bush. Soft contact grunts communicate mutual location among baboons foraging a few yards from each other in dense vegetation. When scanning the surroundings in high grass, monkeys and apes routinely stand on their legs in an upright posture, which they can do for many seconds. The adult males of baboons and of most forest monkeys utter loud calls in response to other parties and groups of their species. Since the calls are highly specific for the species, they carry information on who is where and thus maintain intergroup contact.

GROUP COMPOSITION

The major single variable of group composition in primates is the number of adult males per group. There are two basic group types in this respect. Vervets, macaques, most baboons, most langurs, and most South American species live in multi-male groups. Their social structure permits any number of males per group, even just one. In contrast, species with one-male groups are quite rigid in including one and only one male in each group of the entire population. One-male groups occur at both extremes of the ecological scale, among the extremely arboreal, forest-living guenons and among the extremely terrestrial open-country species, the patas, the gelada, and the hamadryas. It is as yet unknown what ecological factors may be responsible for the one-male groups of the forest species.

The strict limitation to a single group male in these species suggests that their males are highly intolerant of each other at least in the presence of females. Supernumerary males are probably expelled from the heterosexual groups. Patas group males, for example, chase extra-group males away from the group.

Extra males can either lead a solitary life, as among forest guenons or the forest-living but terrestrial drills, or they can join forces and form so-called all-male groups, as among the gelada or the patas. As Struhsacker points out, all-male groups are typical of open-country species, whereas solitary males are typical of the forest. This tendency, which parallels the overall trend toward larger groups in open habitats, may be an adaptation to higher predator pressure or better visibility in the open. The predator hypothesis is unlikely in a species like the patas, which can afford to protect the reproductive groups by only a single male. In their relationship with the heterosexual groups, the "supernumerary" males range anywhere between total isolation to a merely peripheral position within the group.

One possible ecological effect of one-male groups has already been mentioned: With only one male in the group food competition is minimized for the females. In many other respects, the adaptive significance of the one-male group may merely be its small size. A further function appears in the hamadryas baboon where, as we shall see, the male is highly possessive of his females. When we trapped adult hamadryas females for our field experiments, we regularly provoked furious threats and mock attacks by one particular male of the troop. Such a single male, who obviously was the "possessor" of the trapped female, would follow us at about 7 yards' distance when we removed the female. All other males of the troop, though excited, kept a safe distance of about 20 yards. Female anubis, on the other hand, are not associated with any particular male; and they were hardly ever "defended" when trapped. The usual response of the anubis males was to withdraw with the group and watch from a distance as the female was carried away. Thus, the hamadryas female will have only one male protector when her group forages by itself, but most likely one who will defend her with unusual ferocity.

In the past ten years, ethologists have sometimes discussed whether a high level of intra-specific aggression has any causal connection with hunting and killing members of other species, or with active defense against predators. The correlations are, in general, not convincing. In the case of the two baboon species, however, a correlation between social defense and anti-predator defense is probable. The hamadryas male viciously fights any other male who tries to appropriate his female, while anubis males rarely fight over females. A male hamadryas would allow himself to lose his band rather than abandon his female when she is unable to follow because of wounds or disease; in contrast, anubis baboons, at least in Nairobi Park, do not wait for sick group members. In this example, the stronger social bond seems to work against both social rivals and human predators.

There is some evidence that in multi-male groups the number of adult males relative to adult females is affected by the type of habitat. Population samples of anubis baboons, vervets (Cercopithecus aethiops), and Indian langurs (Presbytis entellus) independently show a tendency toward higher proportions of adult females in populations that live under conditions of intense seasonal variation of food availability. It is not known how this ratio is achieved, but the trend may be viewed as adaptive, in that a population exposed to periodic food shortage will depend on a high reproductive potential of its adults to replace seasonal losses. This could be achieved by increasing the number of breeding females relative to the number of males.

ADAPTIVE SPATIAL ARRANGEMENTS

Spatial Arrangements Within the Group

A primate group can alternatively scatter and congregate. According to the Altmanns, the yellow baboons in Amboseli aggregate under the following circumstances: (1) during encounters with potential predators; (2) in response to strong predator alarm calls from another group of baboons; (3) during false alarms; (4) when closely approached by another baboon group or by Masai cattle; (5) in areas of heavy undergrowth; (6) before going through a critical pass in the foliage; (7) when on an unfamiliar route; (8) in response to a spatially restricted resource, such as a water source or the shade of a tree; (9) at or slightly before the beginning of a group progression; (10) in the evening, just before ascending a grove of sleeping-trees; (11) during the morning and evening "social hours." The first seven situations are actually or potentially threatening to the group. Numbers 9 and 10 are times of critical "decision-making." In contrast, the groups disperse in open, safe, and familiar terrain.

In critical situations, terrestrial primates may arrange themselves in such a way that females and juveniles are separated from a potential danger by a shield of adult and subadult males. Troops of geladas often follow the edge of a precipice when they travel. They respond to danger by leaving the plateau and rushing down into the cliffs. During normal progression, the females and juveniles keep closest to the cliff edge, while the males form a screen toward the inland from which dogs and people may approach. (Figure 3.6) In the open savanna of Nairobi park, however, traveling anubis groups are surrounded by peripheral males in every direction. The forest-dwelling and presumably safer anubis groups observed in Uganda do not show such protective progression orders.

En route, females of a hamadryas one-male group typically form a line directed from their own male toward the nearest one-male group, thus leaving the males at the outskirts. However, when two males start to threaten or fight each other, the order reverts to a centrifugal arrangement: The females also line up in the males' "danger shadow," but now away from the fighters in the center.

All such particular orders are restricted to tense situations, social or ecological. The relaxed primate group arranges itself according to "personal friendships" and to the available food, not according to classes of sex and age.

Spatial Arrangements between Groups

The typical primate group keeps aloof from its neighbors and is tied to a particular range. The function of this arrangement is an ordered distribution of resource areas. In addition, the group that remains in its place allows its members to become thoroughly acquainted with the topography of resources and dangers. Such an arrangement in space poses three tasks: to keep the members of each group together; to separate the groups; and to tie each group to a piece of land. Primate groups seem to keep apart and in their respective ranges by various combinations of two mechanisms, which may be called spatial and social. The spatial mechanism rests on the fact that primates, like many other animals, tend to become uneasy when they enter unfamiliar ground. Of the howler monkeys at Barro, Colorado, Carpenter wrote in his pioneer study of 1934: "The location of food and lodge trees as well as prominent arboreal routes within the territory of a group are learned and become positively attractive to the animals. When a clan [i.e., a group] moves toward the border of its territory and the pathway and goals become less and less well-known, there is much frustration, progression is slowed, and the group becomes re-oriented toward the more familiar pathways and better known goals."

A similar tendency is shown by hamadryas bands when they try to enter an unfamiliar sleeping-cliff. If they find the cliff occupied by even a few incompatible residents, they will withdraw even when they are far superior in number. Near their familiar cliff, hamadryas allow a human to approach to within about 50 yards. The same person is avoided at 100 and more yards as the troop enters less and less familiar grounds. Thus, unfamiliar border zones are avoided; but if they are entered, the animals shrink back from encounters.

One may now argue that the repelling quality of unfamiliar border zones is ineffective when home ranges are very small, as is the rule in rich evergreen forests. With a range of less than a square mile, each group can thoroughly familiarize itself with an area much larger than the range necessary to support it. If the group is very small and needs only a fraction of the area which it could cover, the effect will be even more drastic. It is exactly in such small, forest-living groups that one commonly finds the social mechanism of intergroup spacing called territorial behavior. A territory is defined as an area which an individual or group defends against other members of its species, thus obtaining the exclusive use of its resources. Gibbons and the South American Callicebus, both arboreal, are notable examples (Fig. 3.9). Their groups include only one adult male and one adult female. There is evidence that the small one-male groups of

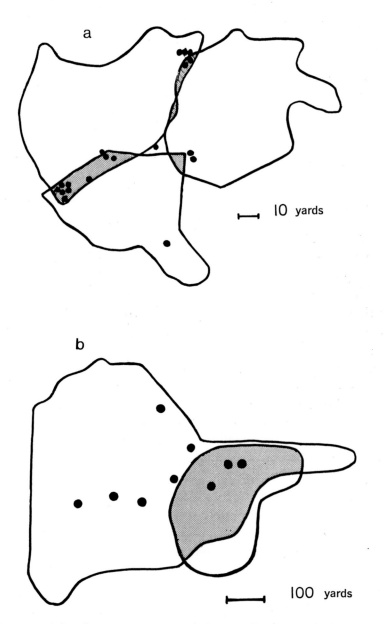

Fig. 3.9. (a) Adjacent territories of three Callicebus moloch groups. Hostile encounters (●) between groups occur mostly near the narrow overlap zones (shaded) of the two territories. (Reproduced from Mason, 1968.) (b) Heavily overlapping home ranges of two langur groups (Presbytis entellus) in Ceylon. Aggressive encounters between groups are not restricted to overlap zones and probably have no territorial function. (Reproduced from Ripley, 1967.)

many arboreal species of Cercopithecus may also be territorial.

In territorial populations, the groups meet one another almost daily near their common borders, sometimes at regular hours and locations. Since they meet familiar neighbors on familiar ground, the general tendency to avoid the unknown does nothing to separate and restrict them. Avoidance comes only as a result of aggression which, however, is mostly ritualized into harmless forms. The animals, especially the adult males, wildly dash through the branches, shaking and breaking them, and there is much calling. In some species these displays occur even when the groups are far apart, thus indicating the groups' locations.

My assumption is that territorial behavior is partly or originally caused by home ranges that are so small that even the neighboring ranges and groups become familiar and thus attractive. It is supported first by psychological data showing that attraction usually increases with familiarity. In addition, there are two species of primates which perform certain elements of territorial behavior, but only in areas where their home ranges are unusually small. The grey langurs (Presbytis entellus) of India and Ceylon have been studied in different and widely separated habitats, ranging from moist deciduous forest to dry and scrub forest with clearings. The considerable differences in social organization between the populations included such variables as home range size, relationships between groups, intensity of dominance behavior, and the relationship between infants and adult males. Table 3.1 reproduces the original table of Yoshiba's (1968) comparative report. The reader is invited to study it carefully and then to attempt hypotheses on the adaptive functions of the behavioral variants in the three habitats. The experience will remind him that even the best data relating to primate socioecology are still tenuous and general. At present we can only interpret some suggestive correlations while many others remain obscure. In the langur example, it is interesting to note that in the two populations with

Table 3.1. Differences of habitat and social behavior between three populations of the Common Indian Langur, Presbytis entellus: Orcha (Eastern Central India), Kaukori (North India), and Dharwar (Western India). (From Yoshiba 1968.)

Characteristics	Orcha	Kaukori	Dharwar
Summer conditions	Moderate	Severe	Severe
Winter conditions	Moderate	Severe	Moderate
Annual rainfall	80 in.; 75% in monsoon	30–50 in.; 70–80% in monsoon	30–50 in.; 90% in monsoon
Natural vegetation	Moist deciduous forest	Dry scrub forest	Dry deciduous forest
Human influence	Very weak	Very strong	Rather strong
Other wild animals	Tiger, leopard, and so forth, abundant	Almost none left	Tiger and so forth, survive in decreased number
Langur population density	7–16 per sq. mi.	7 per sq. mi.	220–349 per sq. mi.
Troop size	22 (average)	54	16 (average)
One-male troop	Less common	–	Common
Nontroop male	Very few	A few	Many
Sex ratio of adult troop members	6 females to a male	3 females to a male	6 females to a male
Home range of a troop	1.5 sq. mi. Seasonal change of core areas	3 sq. mi.	0.072 sq. mi. No seasonal change of core areas
Percent of time on ground per day (approx.)	30–50%	70–80%	20–40%
Weaning age	11–15 months		20 months
Infant male/adult male relations	More tense with less contact and characteristic approach		More relaxed
Juvenile male/adult male relations	More tense with characteristic approach		More relaxed
Subadult male's position in troop social life	Extremely marginal with less contact with the adult male		Near that of the adult male with more contact
Harassment of sexual behavior	By adult and subadult males of the troop		By females of the troop or nontroop males
Male dominance hierarchy	Clearly defined and constant among the adults and subadults of the troop		Not clear between the adult and the subadult of the troop; defined but unstable among nontroop adults

Table 3.1. Continued.

Characteristics	Orcha	Kaukori	Dharwar
Female dominance hierarchy	Observed but poorly defined		Seldom observed
Intertroop relations	Peaceful and tolerant with less frequent encounter		More aggressive with frequent encounters
Relation between the troop and nontroop males	Very aggressive; nontroop males more easily expelled		Very aggressive, with occasional success of nontroop males in entering the troop
Frequency of major social changes		Low	Very high

larger home ranges (1.5 square miles at Orcha, 3 square miles at Kaukori, India), relations between neighboring groups were "peaceful and tolerant," with less frequent encounters. In the two areas with very small home ranges (0.07 square miles at Dharwar, India, about 0.3 square miles at Polonnaruwa, Ceylon), group encounters were frequent and aggressive, involving prolonged chasing, grabbing, and sometimes true fighting. In one of these populations (Ceylon), chases may continue across several home ranges (Fig. 3.9). This supposes that the chasers are indeed familiar with an area that is greater than their home range.

Two populations of vervet monkeys (Cercopithecus aethiops) show the same correlation between intergroup fighting and small home ranges. On ecologically rich Lolui Island in Lake Victoria, population density is high, and the groups forage in fairly small home ranges that are well defined and strongly defended. In the poor habitat at Chobi, north of the lake, where home ranges are larger, intergroup fighting was not reported (Gartlan and Brain 1968).

The ecological function of true territorial behavior is included in its definition as defense of an area. Ellefson (1968) studied group encounters of gibbons in detail and concluded that these encounters indeed have a defensive function. Gibbon territories overlap much less extensively than the home ranges of terrestrial old-world monkeys, with overlap zones only 25 to 80 yards wide. Significantly, intergroup conflicts, which include chasing and biting, occur only in the overlap zones. A gibbon penetrates at most 110 yards into a neighboring territory; he does so only when chasing his neighbor, and hastily retreats thereafter. Many or most of the conflicts arise over a rich food source in the overlap zone. Ellefson argues that the gibbon's food resources are not so abundant as the rich vegetation of the forest would suggest. At any given time, only a few trees in the territory actually carry gibbon food. When gibbons fight, they defend their exclusive access to the food resources of their group range, and are therefore truly territorial. Since the form of their intergroup conflicts is quite rigidly determined, Ellefson concludes that territorial behavior is deeply rooted in the phylogeny of the gibbon genus. The titi monkey, Callicebus moloch, taxonomically quite unrelated to gibbons, has independently evolved nearly the same syndrome.

The aggressive group encounters of the langur populations mentioned above are certainly not deeply rooted phylogenetic adaptations, since they can be present or absent in the same species. They are probably modifications induced by local environmental conditions, such as the extent of the home range. Although langur conflicts look like territorial encounters, their function is not clear. The fights occur not only in overlap zones as in gibbons, but also near the core of a home range. The following field protocol by Suzanne Ripley (1967) shows how a chase between groups can cross several home ranges, which is difficult to interpret as a territorial fight.

April 30, 1963

0810   Beta male and the new subadult male of Troop A run from the middle of their range to the border. They are chasing a group of five males and one female who had been in their range. Then Beta, Gamma, and both the new subadult males chase the group into Troop B_1's range. The strange attack party continues across into Troop C's range. A display whoop is heard from the direction of Troop B. Gamma male from Troop A chases the intruders through Troop B_1's range and to the border of the ranges of troops B_1 and C_1. He then takes a position of good vantage atop a pillar, stares in the direction the intruders have gone, and gives loud grunt threats.

0900   Troop B and C_1 males fight on the border of the ranges of troops B_1 and C_1. Then Troop C_1 males chase Troop B precipitately into Troop B's range. All rest for a short time. Then Troop B_1 males join the fray by whooping close by. Troop B males then continue their retreat well into Troop A's range, even though Troop C_1 has temporarily stopped chasing them. The rest of Troop B_1 cannot be seen.

Suzanne Ripley, who analyzed 31 such group encounters in the Ceylon population, suggests that "these langurs are in some way defending the social integrity of the troop," not a piece of land. Their fights would in this case serve a function which in larger home ranges is fulfilled by great distances between groups. The logical differentiation between defending a territory and defending group integrity may not be self-evident. Its meaning is best illustrated by defining the situation that provokes the "defensive" fight, say, of the dominant group male. In territorial defense, he will attack an outsider whom he sees inside the group's territory, regardless of where the rest of the group is at the moment. In group defense, he will fly into a rage when he sees the trespasser mingling with his group, regardless of whether this happens in the group's home range or somewhere else.

The fights among hamadryas males are a good example of group defense. The male attacks when he sees one of the females of his one-male group sitting too close to another male. The function of the attack is to separate the two, not to remove the other male from a fixed territory. When a one-male group is trapped and released outside its home range near another troop, the group male acts in exactly the same way, showing that his fight is not related to a place but to group membership. Group defense in hamadryas can be artificially provoked by feeding the troop on a small area and thus reducing the distance between one-male groups. Similarly the intergroup fights of the Ceylon langurs might be an effect of their unusual proximity.

True territorial behavior and the avoidance generated by unfamiliar border zones may be assisted by a third and purely social mechanism, hinted at by observations in captivity. It is well known that intra-specific aggression can be directed at an object other than the one that aroused it. A monkey attacked by a more dominant group member very often proceeds to attack a member inferior to himself. It seems likely that aggression originally generated within the group is sometimes redirected against other groups, that is, against strangers. For example, two captive gelada females newly introduced to each other itermittently fought for three days, unable to determine which of them was to assume the dominant position. A number of juveniles were then released into the enclosure. The two females immediately threatened the newcomers, with whom they were even less familiar than with each other. After a while, the two females joined forces; within 15 minutes they groomed each other for the first time, and they did not fight again as long as the youngsters were present. Such redirected aggression may at the same time strengthen cohesion within the group and further the separation among groups.

COOPERATIVE MOTHERING

The primate infant is mainly cared for by its mother. She nurses it and carries it at her belly as it clings to the hair of her sides with hands and feet. Whenever the group is threatened from the outside or when quarrels arise within the group, the mother rushes to pick up the infant. When the infant squeals during play, she threatens or chases its playmates. The younger the infant, the more commonly the mother carries it, even on long walks. The baboon mother carries her young infant at her belly where she can support it with her arm if necessary, whereas older infants ride on the mother's back. The carrying technique seems to differ according to the population. The patas mothers of Uganda have never been observed to carry infants except at their bellies, whereas West African patas also use the back-riding technique. The primate mother defends her infant against group members, but it is the group males who defend it gainst predators.

In several primate species, the mother is assisted by other group members, especially juvenile females and adult females that happen to be without infants. Attracted to the infant, they frequently visit with the mother to watch and sniff the baby. When they find it away from its mother, they pick it up, play with it, or attempt to groom it. Whether the mother tolerates these interactions depends very much on the species. Among langurs, infants are casually passed around soon after birth, whereas mothers of baboons, macaques, and patas are more likely to retrieve the infant or threaten the female that touches it. The interested females therefore resort to rather subtle techniques. A patas female may begin to groom a mother's arm and then slowly and cautiously transfer her grooming to the infant within it. In a captive colony of anubis baboons, females were so successful with such techniques that nearly every mother finally allowed one female to hold and groom her baby. This role which ethologists have unfortunately named the aunt role, usually fell to the mother's closest grooming companion. Some mothers came to use the "aunt" as a baby-sitter while they themselves were feeding elsewhere. In one case, a mother died and the aunt adopted the baby. Among chimpanzees, the infant's older sisters are especially eager to handle it.

Foster mothers are found, however, not only among females. Many adult males also show casual to intense interest in infants. Adult hamadryas males are commonly seen carrying infants within their back manes en route. During rest periods, infants actively seek out certain young adult males who then hug and fondle them. Males, too, can adopt an infant. A motherless hamadryas infant is usually taken over by a young adult male that as yet owns no females. He then carries it en route (Fig. 3.10), allows it to sleep huddled

Fig. 3.10. A young adult hamadryas male carrying a motherless infant that he has adopted.

against his belly at night, and prevents it from moving too far away. Hamadryas males are especially motivated to adopt infants, since adopting juvenile female is their first step toward having a harem of their own.

The survival function of adopting a motherless infant is obvious but the behavior has less conspicuous social advantages for the adopter himself. In some species, such as the Japanese macaque (Macaca fuscata), low-ranking adult males seem to use the attractive qualities of an infant to promote their own social status: By carrying a baby, they are more likely to be admitted into the company of dominant adults.

ADAPTATIONS TO TERRAIN AND CLIMATE

Food, roosting-sites, and predators are the ecological problems most commonly discussed in primate field reports. Here we shall touch on some less conspicuous features of the environment, regardless of whether primates deal with them individually or socially. A fair idea of the limits of primate social adaptations can be obtained only if one also considers what primates have not achieved.

The gelada monkeys of the Ethopian mountains sleep and take refuge in vertical cliffs which can be several hundred yards high. This means that a fall will very probably be fatal. I have heard two reports from reliable observers who saw a fighting gelada male fall from a cliff into the abyss and disappear. My own observations suggest that such accidents may be limited by an inhibition against intense fighting in exposed places. In the field enclosure at the Delta Primate Center, gelada males that came directly from the wild hesitated to fight in the high pine trees. In contrast, a zoo raised gelada male started a fight in a tree 20 yards above ground, fell, and broke a leg. His feral but superior opponent had tried to avoid the fight in the tree and survived the fall unharmed.

Slopes with loose stones are frequent in hamadryas habitats. Quite often such slopes are situated directly above the sleeping-cliffs. Hamadryas baboons immediately respond to the sound of falling stones and skillfully avoid getting hit. An adult hamadryas male was seen to pick up a rolling stone and to hold it in his hands until a group of juveniles playing just underneath him had moved on. Then he dropped it and the stone bounced onto the deserted playground. On the other hand, baboons do not deliberately try to avoid loosening stones by walking more carefully where the ground is loose. This may have given rise to the unconfirmed accounts of baboons throwing stones at humans approaching from below.

Caves are not regularly used for shelter by any known primate population, except for some South African Chacma baboons whose caves are situated in inaccessible cliffs. Caves in flat terrain would be safe only if primates were able to ward off predators, for example with fire; otherwise a cave could easily function as a trap. Cliff ledges protected from above by rocky overhangs are nevertheless preferred sleeping sites of baboons, and they are often sought out when a rainstorm begins. The animals appreciate shelter, but they avoid a place with only one narrow exit.

Rivers and streams of more than a little depth are geographical barriers to most primates. The great apes cannot swim at all; most species have never been observed to swim, with the exception of the genus Macaca, of which several species may swim and dive, both in play and seeking food. Hamadryas baboons cross rivers up to one foot deep, either by carefully selecting a curved path of lesser depth or by a series of long jumps. Mothers have no special way of carrying their infants through water, on the back or the nape instead of under the belly, and this means that the infants are completely immersed after every jump. In many areas, crocodiles make drinking from turbid rivers rather risky. Hamadryas troops that have access to such rivers usually drink only in places protected by boulders separating the shore from the deeper water.

Many forest species do not drink water at all, finding the rain-water on leaves and the water content of their food sufficient for their needs. Desert baboons, at the other extreme, eat dehydrated food for long periods of the year. They depend on drinking water once every day. In the dry seasons, even the remaining waterholes in the sandy river beds turn warm and become infested with algae. When this happens, hamadryas baboons often dig for fresher water in the sand of the river beds, both beside the open pools and in stretches of the river entirely void of surface water. The water that collects in these holes, which are about a foot deep, is cool and clean (Fig. 3.11). The digging is usually done by the larger animals, who then abandon the constantly refilling holes so the youngsters can drink. The anubis baboons around Lake Langano in southern Ethiopia dig similar seepage holes in the shore.

Heavy rain has an immobilizing effect on most primates. Their typical response is to sit quietly in the downpour with heads low. Arboreal primates often remain seated in the foliage of their sleeping-trees. Baboons turn their backs against the wind, with the smaller animals huddling against the chests of larger ones.

Our hamadryas baboons ran for the sheltered ledges in their sleeping-cliff only when they were less than 110 yards away at the onset of rain. Otherwise they made no attempt to select dry spots under rocks or bushes. Mountain gorillas occasionally seek shelter under leaning trunks at the onset of a heavy rainstorm, but they do not move farther than 25 yards from the place where the storm hits them. The nests built by the great apes rarely offer any protection against rain. Jane van Lawick only once observed a chimpanzee nest construction including a roof; once constructed, this nest was repeatedly used by the builder during rain in the following weeks.

Fig. 3.11. Drinking hole dug by hamadryas baboons in the sand of a superficially dry river bed.

Hail is a common phenomenon to primate species living at high altitudes, such as the mountain gorilla or the gelada. Even under such solid precipitation, neither seeks shelter regularly. We saw a large troop of geladas feeding in scattered formation on a grassy slope above the escarpment of the Semien mountains, when fairly large hailstones began to fall. Instead of racing down to the nearby cliffs, the geladas within a few seconds formed tight huddling packs of three to eight individuals until the hail was over. As far as I could see, each pack contained one adult male and probably consisted of one family group. A student of gelada social structure might come to think of hail as rather nice weather.

Primates and snow are, in general, mutually exclusive, although Japanese monkeys and several other species survive temporary covers of snow. These species, and experiences with primates kept in northern zoos, suggest that it is not so much low temperature that limits primates to the warmer climates, but the seasonal shortage of food associated with deep, lasting snow covers and barren trees. That primates have failed to evolve the food storing common to so many lower animals is most surprising and most decisive for their geographical limitation in these areas of winter snows.

Hamadryas baboons avoid high wind by moving to leeward slopes. In cold weather, when their faces and rumps, usually brilliant red, turn bluish, they huddle together and sit quietly, with lowered heads, arms between knees and belly, each animal with its back toward the outside. The preferred huddling unit again is the one-male group, which accomodates most males and all females. Subadult males, who are not allowed close contact with females, often have to sit alone (Fig. 3.12).

This short review of what primates have achieved in dealing with some simple ecological facts should sober anyone who holds that the relatives of man generally excel among animals. There is hardly any technique for handling the physical environment where the primate order is so specialized and advanced as that, say, of the carnivores. In the

Fig. 3.12. Hamadryas baboons huddling in one-male groups on a cool morning. The subadult male "follower" on the left is associated with the group in the foreground but excluded from the huddling.

final chapter, we shall consider the puzzling question of why human technology grew from a stock of very untechnical animals.

SUMMARY

1. A primate group is best defined as a number of individuals who remain spatially together in the population and who interact more frequently with each other than with outsiders.

2. An ideal foraging group can be expected to contain as many animals as can simultaneously feed on one resource unit. The correlation will be most critical with respect to vital resource units situated far apart. Species living in harsh environments with resource units of very different size tend to adopt a flexible fusion-fission society.

3. Large groups are better at discovering, mobbing, or chasing a predator than small ones. Resource conditions seem to prevent large groups in the patas monkey; a solution is found in a watchful male and a tendency among the females to hide, combined with great running speed. In judging the adaptive value of group size, it is necessary to examine each function of the group separately.

4. Behaviors that are favorably or necessarily carried out by all group members at the same time are subject to social facilitation. Social inhibitions seem to discourage behaviors that are preferably carried out by one or a few members only.

5. The activity of a dominant animal inhibits the same activity in an inferior neighbor, and thus fights are avoided. Feeding in trees compensates for the disadvantage of inferior animals in that they may be supported by thinner branches. Dominance keeps feeding animals at a distance from each other. Individuals of high status may attract much attention from others and thus are potential leaders. However, not all social roles are necessarily correlated with dominance.

6. The importance of planned travel routes and leadership increases with the distances to be covered. Primate groups are generally not led by a single permanent leader, except in some species with one-male groups. Group members may tend in markedly different directions, but they do not quarrel in such situations. Exploring parties are rarely used. In two baboon species, initiative and decision appear as distinct roles. Locations associated with negative experiences can be avoided. Cohesion in dense vegetation is supported by vocalizations.

7. Territorial defense is a deeply rooted phylogenetic adaptation in Callicebus monkeys and gibbons. Aggressive encounters between groups are ritualized and occur only near the common border. In langurs intergroup fighting is observed as a modification in areas with very small home ranges. Here, fights are less ritualized and not restricted to the border zone. Familiarity of the terrain and redirected aggression may affect the tendency for fighting between groups.

8. A review of the ecological techniques of the individual primate shows that primates do not shape their habitats, with the exception of modest achievements among the apes. This raises the question of why it was the primate order that finally evolved the most technical animal.

Chapter 4

WAYS OF ADAPTING

THE CONTRIBUTION OF THE CAUSAL VIEWPOINT

So far we have discussed the rewards of group life; we must also examine the costs. Finding that a given type of social structure is adaptive under a given set of conditions does not explain how that social structure was or could be realized. Our own species sufficiently demonstrates that not everything that is adaptive is also possible. Cultural developments, like other modifications, are restricted by the phylogenetic heritage of the species, and phylogenetic adaption itself proceeds slowly. Ecological conditions can put a premium on a certain type of society, but it cannot tell the species how to create such a society. Discussions of adaptiveness sometimes leave us with the impression that every trait observed in a species must by definition be ideally adaptive, whereas all we can say with certainty is that it must be tolerable since it did not lead to extinction. Evolution, after all, is not sorcery.

Chimpanzees, for example, live in an open society in which large, well-defined groups have not been identified. The parties of a wide area meet and recombine in a constant process of reorganization. This seems to be an ecologically ideal fusion-fission system, providing group sizes and compositions for every conceivable context. Surprisingly, it has not been found in even one species of monkey. The conclusion that an open society is maladaptive for all monkeys, however, is unacceptable. It is quite possible that monkeys would benefit from such societies but do not form them because they are unable to do so with their social dispositions. The question can be decided only by research on the *causation* of social organization.

Only a combination of the functional and the causal viewpoints can show whether a trait of society is optimally adaptive or whether it is merely the suboptimal best for the population with its inherited and traditional stock of social techniques. While the first part of this book was devoted to the functions of societies, we shall now concentrate on their causation.

When a new or changing environment requires of a species an adaptation of one of its features, several solutions may be feasible. But the immigrant or surviving population is probably not capable of realizing every one of these solutions within a tolerable time span. A solution arises partly from long-standing predispositions and partly from adaptations to a former habitat. If a species now succeeds in adapting by introducing a structural change compatible with its preexisting structure, it thereby alters a component of its social, behavioral, or morphological organization. The result may be disharmony within the organization, so that other components must be altered to accommodate the main change and to reestablish a functioning entity. Such secondary adjustments also must be within the capacity of the heritage, and the new environment may then permit only some of them.

Recording a case of adaptive behavior is therefore not enough. An equally interesting task is to understand how a species recombined, extended, and reduced the elements and subsystems that made up its previous adaptation, how it built its new life on the limitations and assets of its inherited potential. With this additional viewpoint, the anthropologist's interest in primates will not be merely comparative; it will help him to think about which parts of the primate heritage early hominids brought with them and about what they did and what they could not do with them in their subsequent history.

In this chapter, we shall first describe an example of a primate society that probably adapted phylogenetically. The behavioral means developed in the process will then be compared with alternative means of similarly constructed societies, leading to a summary list of the social behaviors that seem to shape primate societies. The third section will present an example of adaptive modification by tradition, and, in Chapter 5, we shall examine a preliminary method of distinguishing between the two types of adaptation.

ADAPTATION BY EVOLUTION

Hamadryas: A Phylogenetic Adaptation Reconstructed

The hamadryas baboon may serve as an example of a society's probable phylogenetic change. Since its closest relatives are alive and known, its evolutionary pathway of adaptation is not too difficult to visualize.

In terms of numbers and distribution, baboons are a successful primate genus. Their five species are found over wide areas of Africa, in five distinct ranges varying from forest to semi-desert. Morphologically, baboons are quadrupedal walkers, spending much time on the ground. They forage in the savanna grass for roots, seeds, leaves, and insects, and also in trees for blossoms and fruits; only rarely do they kill and eat small gazelles. Water must be found daily. Their life in the open country makes them accessible to large carnivores, and since they cannot run very fast or far, their safety sometimes lies in the protectiveness of their large males. Their nights, however, are spent in the relative safety of high trees or, in some areas, of vertical cliffs.

Baboon social organization is quite similar throughout the continent. The typical baboon population is divided into groups that tend to avoid each other without territorial disputes. When a group, which varies in number from 10 to about 100 animals, occupies an area on one edge of its home range, the neighboring group on that side tends to move away. Not all groups are equally tolerant of each other. Some may quietly drink side by side at the same waterhole, whereas others act nervous in a similar situation; the adult males of each group may then cluster at the points where the groups are closest. Fighting between two groups was observed when both tried to settle in the same sleeping-grove.

There are no lasting sub-units within the group except for the bond between a mother and her infant. When a female reaches the ovulation period (oestrus) of her monthly cycle, she associates with a male for mating and frequent grooming. Usually the most dominant among the interested males obtains the female, but several males may succeed each other as consorts of a female during a single oestrus. Typically, the consort male follows his female whenever she moves a few steps. In a few groups of anubis baboons, the consort pair is followed around and harassed by rival males who seek to obtain the female; in other areas, the consort male maintains his exclusive access to the female without quarrel. The male and the female separate again after the period of her oestrus, and both resume their usual contacts with various group members.

Only one of the five species, the hamadryas, deviates from the general social structure of baboons. Every hamadryas female is permanently associated with a particular male, regardless whether or not she is in oestrus or pregnant. Since one hamadryas male owns several females, he cannot possibly follow them all; instead he forces them to follow him by attacking them when they go too far away. Several such one-male groups and some single males are joined in a band about the size of an anubis group. Two or more bands in turn form a troop.

How did the two social structures of the baboon genus differentiate? The closest relatives of baboons, the macaques, are organized in the general baboon pattern; one-male groups have not been found consistently in any macaque species, but consort pairs are common. Since the hamadryas are the only deviators, and since they live in an extreme environment compared to that of other primates, it is most likely that their organization is a recent specialization for their arid habitat. At one time, their ancestors probably shared the social ways of the other baboon species in a more favorable habitat. This is the hypothesis that we shall now follow in comparing the hamadryas with their immediate geographical neighbors, the anubis baboons (Figs. 4.1, 4.2).

When the ancestors of present-day hamadryas began to adapt to their new semi-desert environment, they were not just nondescript animals ready for any way of life; they were baboons. They were built to walk the ground on all fours for long distances rather than spend their days swinging in the trees. They avoided predators by sharp eyesight and by running if necessary, not by hiding. They were organized into groups of up to a hundred, relying on each others' support in discouraging cheetahs and leopards. Groups avoided each other, though a single stranger could join a troop and find his way into its hierarchy. The males were twice the size of their females and quite capable of keeping jackals away from their young. They brought with them a pronounced dominance order in which the low-ranking yielded to the higher, and they chased each other frequently and with much noise. Their males had no permanent female consorts, but paired only when the female was in oestrus and sexually receptive.

This was the raw material from which the adaptation to desert life had to be built. Any drastic deviation from the heritage was unlikely. The ancestral hamadryas might have changed the size of their groups, but they would not easily turn to a solitary life. They could perhaps be expected to reorganize their society, but not to abandon dominance in return for food-sharing and an open society. The incipient desert dwellers were not just baboons; they were higher primates. They might have developed an ability to run faster,

Fig. 4.1. Habitat of anubis baboons (Papio anubis) above Awash Falls; the sleeping trees in the gallery forest are seen from the open bush country. (Photograph by U. Nagel.)

Fig. 4.2. Anubis baboons foraging near the gallery forest. As among the hamadryas, adult males weigh as much as two females, but anubis males lack the long grey mantle that marks the harem leaders of hamadryas. (Photograph by U. Nagel.)

but not to dig burrows for protection. Nocturnal life had long ago been abandoned by their ancestors, so their existing sensory and behavioral equipment would have made it difficult for them to revert to living in the dark. And finally, being vertebrates, they shared even more basic characteristics of social life; they knew each other as individuals and were used to competing for individual positions and tasks. They would not develop an ant-like society, where individuality was neither important nor recognized among hundreds of conspecifics.

In which respects did the semi-desert require new adaptations? The essential resources—vegetable food, water, and safe elevated places—remained unchanged, though the trees were replaced by cliffs. The exploitation of resources thus could follow the old pattern. Each member of the troop could search for his own food and eat it on the spot, drink water daily, and spend the nights above ground. The adaptations required by the slightly different quality of the resources were minimal.

These adaptations were perhaps realized solely by individual modification and by tradition. Phylogenetic adaptation may have been unnecessary in many respects. This, at least, is suggested by field experiments: Anubis females that were trapped and then released beyond their species border into a hamadryas troop survived for months. They could live on the local food; they were able to drink frequently enough; and although they were awkward rock-climbers and easily slipped when chased at first, they soon learned. Some adaptations from an anubis to a hamadryas environment and vice versa are thus possible within a fraction of one life span, at least if the transferred animal profits from the experience and leadership of its fully adapted host troop. These experiments, of course, do not indicate that the transplants ultimately adapted so well as their hosts. Their chance of survival may have been several percentage points smaller, and such a shortcoming could very well exterminate the transplanted genotype in a few generations. The hamadryas, after all, did change its ways, and the nature of the changes suggests the environmental differences that favored them.

Hamadryas baboons abandoned the one-level society of their fellow baboons and substituted three levels for it: the large troop, the medium-sized band, and the small one-male unit. We have already seen that their semi-desert habitat requires groups of several sizes, according to the function or task at hand. The independent one-male group appears as the optimal foraging unit, just large enough to include one male protector, but small enough to find enough food for every member without covering inordinate distances. (In times of severe food shortage, the groups of the South African chacma baboon also split into temporary one-male groups, although they are normally of the anubis type.) On the other hand, the one-male units must gather at the few available cliffs in numbers far exceeding the group size of non-desert baboons. The social solution is the troop. The band represents an intermediate level between the small foraging and the large sleeping units. Its ecological functions are not immediately evident; the band is quite possibly a vestigial unit for the hamadryas baboon, that is, a remainder of the ancestral baboon group. Hamadryas bands resemble groups of other baboons in that they are closed units rarely interacting with and often wary of each other. In the hamadryas, the baboon heritage succeeded in producing two new types of associations, the troop and the one-male group, and we are now interested in the behavioral innovations and rearrangements that produced them.

Let us first consider the troop. Since its members do not interact except within each band, the troop is not a genuine social unit. Tolerance of proximity among bands was all the troop level required. This ability was already part of the general baboon heritage. In some parts of the Amboseli National Park of Kenya, for example, groves of sleeping-trees are few but large; in this respect the yellow baboons

(Papio cynocephalus) of the area face the same problem as the hamadryas and solve it in the same way; groups tolerate each others' presence in the same roosting-grove. This local adaptation suggests that the hamadryas troop is not a spectacular design with a long and difficult history. It could probably be realized by any baboon population within a rather short period without much or any genetic alteration.

Evolving the one-male group was more complex and probably took much longer. These small units are based on permanent bonds between one or more females and one male. Since infants and small juveniles are in turn bonded to their mothers, the band is fragmented into neat parcels with only some adult and juvenile males left out for variable associations. Each group male regularly checks on his females' whereabouts. On the march, he frequently looks back at them (Fig. 2.5). When a female is too far away, or when she interacts with a band member not belonging to the group, he threatens her by raising his brows, or he attacks and bites her on the back or the nape of the neck. The females respond to such aggression by resuming their positions close to him.

This herding technique, reminiscent of a baboon mother controlling her infant, apparently was the best means available to baboons for a small, stable foraging unit. It is inelegant and consumes a great deal of the male's attention, but it draws on some important preadaptations of the baboon heritage. First, baboon social life is dominance-oriented anyway. Second, the large size of the males makes it easy for them to dominate more than one female. Sexual dimorphism of size is found among many vertebrates where males herd a harem of females. Since all baboons show this dimorphism regardless of their social structure, it probably already existed before the hamadryas began to evolve one-male groups. The third preadaptation is the tendency for exclusive bonding present in all baboons and macaques. This means that an individual aggressively defends his exclusive access to a highly desirable social partner. Thus, baboon mothers are intolerant of other females' handling their infants, and consort males do not allow rivals to mate with their females. The latter is in sharp contrast to the behavior of gorillas and chimpanzees.

In the hamadryas male, a high motivation to associate with females merely had to be extended to the anoestrous female to make him the intolerant permanent mate he now is. To own more than one female, however, required an innovation, the herding technique. The repertoire of baboons already included its behavioral components, that is, brow-raising and neck-biting. The hamadryas, however, are the only baboon species that built these behaviors into a tool for keeping females close by timing their attack to the female's behavior. Anubis males never herd their females even when they consort with them. Even an adult anubis male who lived for several months in a wild hamadryas troop did not adopt the herding technique, and this completely deprived him of female company.

The hamadryas males' possessiveness toward females in every reproductive state may well have a genetic basis. Herding and one-male groups have been found in all wild troops observed so far, although differences in intensity could be noticed. The hamadryas colony at the Russian research station of Sukhumi still shows herding behavior after several generations in captivity. Apparently, even a drastic change of environment does not necessarily alter the herding syndrome.

While the behavioral and morphological dispositions of baboons prepared the way for the one-male group, the new system also required some secondary social adaptations. For one thing, hamadryas females have to respond to a male threat by approaching the aggressor, whereas any other primate female, including anubis, takes flight in the same situation. This paradoxical reversal of the "normal" response stimulated a series of field experiments in which adult anubis females were transferred to hamadryas troops. The results were quite impressive. First, the anubis females were eagerly accepted and herded by resident hamadryas males despite the females' different appearance. Second, the species difference did not prevent the animals from communicating successfully. Furthermore, the anubis females learned, within one hour on the average, to follow the one hamadryas male who would threaten and attack them, and to interact with no other male. In fact, their scores for following reached the same level as the hamadryas control females that were transferred into the same troop.

The experiment demonstrated that to follow and approach a threatening male instead of escaping him is a task that a baboon brain can master without special genetic support. In reverse experiments, hamadryas females readily adapted to the independent life in an anubis group; they groomed several males in succession and ceased to follow any particular male. Whereas male herding appears as an inert, perhaps phylogenetic, adaptation, female behavior is flexible, and therefore required little phylogenetic adaptation, except perhaps for one trait: Transferred anubis females retained a higher escape tendency than hamadryas controls even though they had learned the following-response. Females escaping their males sometimes lose the troop. Since unprotected females may more easily be caught by predators, the selective process that lowered escape tendencies in hamadryas females seems obvious.

On the side of the males, the new social system also introduced a new problem. It created males who, already aggressive were now highly possessive of females (Fig. 4.3). Whereas anubis females may provoke male competition for a few days each month, hamadryas females became permanent incentives. When 30 new females were introduced to a colony of about 100 hamadryas baboons in the London Zoo, all the old males tried to secure females and, within one month, killed 15 of them in competitive fights over their possession. This event, though provoked by an unnatural manipulation, points out the risk in evolving possessive males.

In the course of our studies, my collaborators and I became convinced that hamadryas baboons had evolved an efficient design to counter this risk. We had so far released around 30 adult baboon females into strange hamadryas troops during various experiments. In moderate doses, we thus replicated the London "experiment." The outcome, however, was very different. There never was a general rush of the troop males for the new female. Regularly, only one male of the troop came forward and appropriated the female. A few times two males would approach simultaneously at the first sight of the female, but within seconds, one of them would withdraw. Only once did a short fight between two males break out. This generally restrained behavior could not be explained by a low motivation of free-living males to appropriate females. For one thing, every released female immediately found one interested male. Furthermore, freshly trapped adult hamadryas males in a field enclosure at our camp eagerly took over every new female introduced to them. The possessive motivation sometimes assumed bizarre forms. A hamadryas male was turned loose near his troop after he had served in a series of enclosure experiments. Instead of joining his troop, he raced after the departing truck, trying to join some females whom he knew from the experiments and who were now sitting in cages on its roof rack.

Lack of interest certainly is not the explanation for the lack of aggressive competition in wild troops, nor is a lack of aggressive motivation. Fights over females can easily be produced by the experimenter even in the wild (Fig. 4.4). Whenever a male is trapped and temporarily removed, his females are taken over by other males of his troop. After only a few hours, the new possessor will not release the acquired female without a fight, even if the former leader is returned. The outcome of the fight will then determine whether the female remains with her original possessor or with the new one. This raises an answer and a question: Fighting ability can obviously decide who will own a female, but if it is the new possessor who wins, why did he not attack the original

Fig. 4.3. During an aggressive episode in the troop, a hamadryas male protects his assembled females with a "threat yawn" directed at other troop males.

Fig. 4.4. Artificial feeding of a hamadryas troop leads first to crowding; this, in turn, makes the males uncertain about the integrity of their harems. While rival males threaten each other, their females line up in the "attack shadows" of their respective owners.

one and conquer the female long ago? Fighting power is clearly not the whole answer. If it were, we would expect that a few superior fighters would appropriate all the females of a troop, and this, in reality, is not true. In the Erer-Gota population, no less than 80 per cent of all adult males owned females, and an analysis of a band in 1968 revealed that the number of owned females in a harem did not correlate with the male's dominance position. Weaker males can definitely own females.

These facts led us to postulate a mechanism that moderates the effect of fighting power. We hypothesized that a male does not *claim* a female if she already belongs to another male. To test the hypothesis we designed a simple experiment. Adult hamadryas baboons were trapped, brought to the camp, and habituated to an enclosure situated outside their previous home ranges. This neutral ground was chosen because familiar terrain and the presence of familiar troop members might have affected the males' chances of success. A male and a female were then released in the enclosure, while a second male watched them from a cage 10 yards away. In this experiment, the female was initially a stranger to both of the males, whereas the males came from the same troop and could be expected to know each other. In the enclosure the new possessor at once initiated the mounting, grooming, and following typical of a hamadryas pair bond. The information offered to the onlooking male thus was that a male of his own troop was engaging in pairing activities with a female new to the troop.

When the onlooker was introduced to the pair after 15 minutes of watching, he behaved rather strikingly. He refrained from fighting over the female and clearly avoided even looking at the pair. Most of the time he sat in the corner opposite the others, facing away from them. At intervals, he looked at the sky, inspected the well-known countryside as if something were moving there, or aimlessly fiddled on the ground with one hand. His social behavior was strongly inhibited. Control experiments had shown that the two males without a female readily interacted, and that without the first male's presence the onlooker would not hesitate to appropriate a female. What inhibited him, then, was the two partners combined, the "pair Gestalt." The possessor's social behavior, in contrast, was disinhibited more than normal and included frequent approaches to the onlooker.

This result could possibly still be explained by the superior fighting capacity of the possessor, known to the onlooker from their common life in the same troop. Therefore, the experiment was repeated two days later with reversed roles. The former onlooker now became the possessor of a new female, and the former possessor was introduced after watching the pair. If fighting power had decided the first experiment, the new onlooker should now have appropriated the female. In reality, it was he who now showed all the signs of inhibition. The female again remained with the one who had first established the pair bond with her. In repeated experiments with a total of eight males, the possessor invariably retained the female, regardless of fighting capacity or dominance. Regularly, the "have-nots" were inhibited while "haves" were disinhibited. If these findings are generally true, it becomes clear why males of low fighting power can retain harems in the troops. The inhibiting effect of the pair Gestalt is a stabilizing factor that protects pair bonds and lends the one-male group a certain immunity.

It appears, then, that hamadryas baboons, in evolving a system of one-male units in adaptation to their environment, also evolved a mechanism to cope with the concomitant problem of highly possessive males; possessiveness was complemented by a "respect" for possession. The inhibition against encroaching on another's social property is at least analoguous to certain rules of human behavior, and the fact that such inhibitions occur in nonhuman primates is of great interest in the search for the roots of human social behavior.

The inhibition of hamadryas males restricts fighting over females to exceptional situations, such as extreme crowding. The solution obviously corrects some negative effects of the

harem system. But the change of social structure has probably had other ramifications in the social system of the
hamadryas, requiring other secondary adaptations. For example, with the inhibition it is difficult for young adult males
to acquire any females at all. How this new difficulty—established old males' holding all the group females—was solved
is not clearly understood, except for one characteristic that
obviously suggests a solution. Some young adult males acquire small juvenile females long before the latter are sexually mature. After an initial phase of intense herding and
mothering behavior by the male, the young female follows
him permanently and very probably grows to be his first
mature consort. Acquiring juvenile females is easy for the
young adults because males in full prime are only interested
in adult females. Thus the younger male can bypass the inhibition barrier. By the time his juvenile consorts have
reached adulthood, inhibition works for him, protecting his
now generally attractive females from older males.

The hamadryas society may be of special interest with
respect to man. Independent of each other, man and hamadryas baboons both evolved stable family units linked together in larger bands, and both presumably arrived at this
type of society when living in open country. Anthropologists
have explained the origin of the human family in several
ways. The phylogenetic gap between man and hamadryas
will not permit proving or disproving these explanations on
the basis of hamadryas evolution, but we can at least see
whether they are compatible with what we know from the
baboon.

One hypothesis points out that the permanent sexual
receptivity of the human female may have evolved in support of a stable male-female bond. The attraction between
partners was thus perpetuated, and regular social life was not
upset by the mania of oestrus with its outbursts of sexual
rivalry. From this hypothesis, we should predict that pair
formation within a primate group is more likely when females are sexually attractive than when they are not. Anubis

baboons are a case in support: the consort pairs between a male and a female are formed only when and as long as the female is sexually attractive. Experiments might try to produce permanent consort pairs by inducing continuous sexual attractiveness.

The hamadryas, on the other hand, seems to disprove the hypothesis. Hamadryas females are receptive for the same number of days a month as anubis females, but this short oestrus is a sufficient basis for a permanent consort relationship. However, the hypothesis could easily be saved in slightly more general form. Apparently, the female's attractiveness can be other than sexual. As yet, we do not know what is so attractive in the hamadryas female, but we do know that the attraction is so powerful that periodic sexual stimuli do nothing to increase the male's total interest in the female. We can thus assume that any kind of increased attraction between the sexes might cause permanent pairs. Man and hamadryas would differ only in the quality of intrasexual attraction. Whereas the human female may have changed to become more permanently attractive, in the hamadryas it was the male that changed to become more permanently attracted.

The hypothesis, nevertheless, is still incomplete. Increased attraction will not only produce pairs but may also increase competition, at least if the males have a strong polygynous tendency. As a necessary addition to the hypothesis, we must postulate a mechanism that protects a once-established pair from competition by rival males. By pointing out that such a mechanism has now been discovered in the hamadryas I do not suggest that humans have or should have similar inhibitions! I only suggest that attraction between the sexes seems a logically insufficient causal explanation of stable pair bonds.

Anthropologists further assume that economic factors united the human pair more closely. As food-gathering by both sexes was replaced by male hunting and female gathering, food-sharing between males and females made the members of the pair dependent on each other. In hamadryas baboons, however, the female is forced into a dependence that is not based on the exchange of food. It seems possible that man, too, evolved the pair bond before the division of labor. Food-sharing probably came more easily between a male and his consort female than between a number of males and a number of females. The pair members probably had an especially high tolerance for each other in the first place, each being predominantly attractive to the other. Even in pre-sharing epochs, the pair would certainly have consisted of two individuals that could at least feed near each other without competing.

Behavior Sets That Shape Societies

One may now question whether herding behavior plus an inhibition of male encroaching was the only possibility for evolving one-male groups, or whether other primates reached a similar group type with other behavioral means, involving, for example, a remodeling of female behavior. Fortunately, the one-male group does not occur only among hamadryas baboons. It has been found among several species of forest-living Cercopithecus monkeys and two old-world species, Theropithecus gelada and Erythrocebus patas. The last two, the gelada and the patas, are open-country dwellers like the hamadryas; they resemble baboons in their way of life and general appearance. The males are twice as large as the females, and the basic social unit is a group consisting of one adult male and several females. These two species and the hamadryas probably evolved their one-male groups under similar ecological pressures; but as we shall see, the groups evolved from different syndromes of social behavior.

The gelada, although it is not a Papio, is often called the "gelada baboon." It lives in large troops on the open alpine meadows of the Ethiopian highlands at altitudes up to 4,400 yards. At night the geladas descend to the tremendous cliffs to sleep on ledges or steep grassy patches. During the day, they travel and forage on the treeless meadows above, never far from the safe abyss, and with a protective belt of large males on the inland side of the troop. John Crook, the British psychologist who first studied them in the wild, discovered that the troops were composed of one-male groups which may separate for independent foraging. However, Crook made it quite clear that the female's membership in the one-male group was not enforced by male herding. Gelada females are allowed to disperse freely throughout the troop.

Shortly after Crook's study, we had the opportunity to establish two gelada colonies in a large enclosure at the Delta Primate Research Center in Covington, Louisiana. The animals, introduced to the enclosure as mutual strangers, established an organization of one-male groups and in the process demonstrated the behavioral mechanisms responsible for their social structure. Unlike hamadryas society, the gelada one-male groups were formed through the activities of both sexes. Immediately after the introduction a male would pair off with a dominant female: The female would present her rear, he would mount her, and she would then groom his cape. From this point on, the pair female would viciously attack any other female who approached her male or whom her male sought to acquire in addition to herself. A second female was eventually accepted, but only after the first, dominant female had gone through the pair-forming process with female number two, with herself in the role of the male. Number two had to present her rear, be mounted by female number one, and then groom her. Thus the dominant female established herself in the role of the owner of number two, just as the male had established himself as her owner.

As a result, each one-male group comprised a chain of animals each of whom dominated and controlled an immediate subordinate, preventing its interacting with nearly everyone else in the colony. One top female, for example, interfered with every interaction between her male and the number two female to such an extent that the latter learned to avoid the approaching male. Only when the colonies were well stabilized did contacts between the male and the number two female become more frequent. Even then high-ranking females would drive their subordinates back to the group when they attempted to contact members of another one-male group. The dominant gelada female is, as it were, a secondary group leader. She is usually nearest the male, and in encounters with strange males proves to be the most faithful of his consorts.

The one-male groups of gelada and hamadryas have in common the control and restriction of females by more dominant group members. This role, however, is monopolized by the male in hamadryas baboons; in geladas it is shared by the females. As if in compensation, the herding technique of geladas is poorly developed. Although the males have the same motor repertoire, ranging from threats to neck-bites, they apparently do not apply it consistently enough to condition a reliable following-response in their females. In our colony, none of the females responded to a male threat or attack by following, as a hamadryas female would have done. A male leader had to prevent interactions with outsiders by going to the spot and aggressively separating his female from the stranger. Herding by females followed the same pattern; a dominant female would sometimes have to drag a younger group member back to the group with her hands.

It is not known whether gelada males have a hamadryas type of inhibition with respect to each other's females, but it seems that the gelada system can function without it. To win a female over from another group, a gelada male has to do more than just defeat her owner (which would suffice in hamadryas society). In our gelada colony, attempts to appropriate a female met the solid resistance not only of her male but also of most females in both groups. The claimed female herself often assisted her leader in chasing her new suitor. The integrity of gelada one-male groups rests on the joint action of many. This multiple safeguard may be the reason that a gelada one-male group can spread throughout the troop, whereas the hamadryas harem depends on spatial coherence within the troop.

Patas monkeys form no troops. Agile fleers and hiders, they do not depend on large social units; their one-male groups live apart from each other, so that the male is not forced to live near females that are not his own. Such a social system is much easier to maintain than the crowded aggregations of one-male groups of the gelada or hamadryas. First, herding is not required, and patas males in fact do not do it. The exclusion of strangers is not a permanent task but an occasional event that totally separates the groups for a number of days. All that is required is that males be highly intolerant of each other. In contrast, hamadryas or geladas must constantly maintain the discreteness of their one-male groups, though intolerance among males must be moderated so as not to split the troop.

Among the closest relatives of the patas monkey are the vervets (Cercopithecus aethiops), which are organized into multi-male groups. They live on the edge of the savanna. If the patas descended from ancestors with a similar society, the main behavioral change was an increase of male antagonism. Studies of patas monkeys in captivity amply demonstrate this antagonism. When the young male of a group studied by Hall had reached puberty, the single adult male began to attack him so viciously that the adolescent had to be removed, but he "remained in a state of apparent terror for over an hour, despite his removal from danger." At the Delta Primate Center, we made similar observations on pairs of adult patas males released into a 100-by-400-foot enclosure. The two males would immediately begin to fight for no apparent cause except each other's presence. After several rounds, one of them would emerge as the winner. But no matter how far the weaker male now withdrew, the winner would relentlessly seek him out and chase him every few minutes (Figure 4.5). After an hour or so the fleeing male would begin to scream, a rather unusual behavior for an adult primate male. Eventually, after a day or two, the loser lay passively in a corner and refused to eat except from the hands of the observer, whom he approached as if for protection. Had he not been removed, he probably would have died from stress, although he was not wounded. Vicious, persisting intolerance among adult males emerged as the patas route toward a solitary one-male group.

At this point one begins to wonder why the leader of a gelada or hamadryas harem prefers life in a troop, with its constant risk to the integrity of his one-male group. That troop life is adaptive for them is only the ultimate evolutionary cause, not the immediate motive of the individual male's gregarious behavior. The immediate cause is a high tolerance and attraction among the group leaders, an attraction that outweighs the tendency to avoid potential rivals.

The core of the difference between patas and gelada males emerged when the latter were subjected to the experiment described above. Two gelada males, introduced to the enclosure, would also fight, and a winner would soon emerge. But once the loser ceased to attack and began to avoid his opponent, the winner's behavior changed. He no longer threatened or chased the weaker male but began to approach him with the friendly gesture of lip-smacking. The loser's response was to withdraw or even to attack when cornered, but the winner continued his friendly approaches. Hours later, the loser would nervously present his hindquarters in a gesture of submission. The winner would then cautiously mount him and invite him to groom, until the subordinate male began to run his hands through the winner's hair. On the second day, the loser appeared relaxed; the two males sat and foraged side by side and frequently engaged in social grooming. The behavioral design that permits troop life in geladas thus is submission (Fig. 4.6). It is significant that no submissive gesture is known for the patas monkey.

Fig. 4.5. An adult patas male driving a screaming rival (left). Patas males lack the social technique of submission. (Delta Primate Research Center.)

Patas, geladas, and hamadryas arrived at their one-male groups by convergent evolution. In their organizations, so-called aggression assumes three markedly different functions.

Fig. 4.6. After an initial fight two adult gelada males have established a dominance relationship by the submission of the male in the foreground. They now can walk and forage side by side. (Delta Primate Research Center.)

Among patas males, it separates the males by distance; among geladas and hamadryas, it establishes the compatible and mutually attractive roles of the dominant and the subordinate; and in the herding of the hamadryas, it discourages the females from leaving the male. So-called attraction appears in two forms: A bilateral form that causes both partners to seek each other's proximity, and a one-sided type in which one partner enforces the other's proximity.

This simplified model requires further elaboration. For example, several field workers observed groups of adult and subadult patas males with no females or young. This suggests that there is an attraction even among patas males, which, however, breaks down in the presence of females. Females apparently have the same effect in some other species of primates. In the Dharwar area in India, all-male langur groups are common. Such a group occasionally cooperates in expelling the single male leader of a neighboring one-male group. However, immediately after the takeover, the previously compatible males of the all-male group begin to fight until a single male has expelled all the others and remains in sole possession of the females. The disruptive effect of the female presence can even be discovered among troop-living hamadryas and captive geladas. Here, males having no females sit close to each other and frequently groom one another, but the males who own females keep farther apart in the troop and are never seen to groom each other. In these two species, male-male attraction is strong enough to keep the males in the same troop even though the presence of females counteracts it considerably. Among the hamadryas, only large-scale fights over females succeed in reducing the attraction among males to zero. Under these conditions, the troop splits into the component one-male groups and, for a few hours, a patas-like situation occurs.

The causal investigation of primate societies will have to explain the existence of groups as such. Why is it that a population does not assemble into one enormous mass of individuals? Why is a primate attracted by a member of his own group while he chases or avoids a stranger, though both individuals carry very similar sets of stimuli and seem equally satisfactory as social partners? The astonishing polarization in the treatment of members and outsiders has not yet received attention by field workers, and I shall therefore refrain from tentative explanations.

This chapter has indicated an incipient line of research which, in the long run, will show us the inventory of behavioral systems that primates have used in evolving adaptive social structures. According to the preliminary knowledge now available, primates employ the following behavioral sets in shaping a particular social organization:

1. Class affinities: Members of certain sex and age classes may attract each other, while others, especially members of the same class, may be mutually intolerant.

2. Triadic interference: The dyadic affinities can be mediated or suppressed by the presence of members of a third class.

3. Stranger-member polarization: There is more or less pronounced discrimination between members of the same group, who are generally attractive, and of neighboring groups, who are avoided or attacked. The familiarity of the location in which two groups meet may affect the style of their encounter.

4. Peripheral tendency in males: Males are more or less inclined to a peripheral and solitary existence, whereas females tend to remain with a group and stay near its center.

5. Exclusive bonding: Within the group, certain members more or less intensely defend their exclusive access to certain preferred partners.

6. Facilitation and inhibition: Social facilitation can synchronize particular types of behavior, whereas social inhibition (e.g., dominance) can restrict certain behaviors to carriers of certain roles.

7. Role coercion: A group member can adapt to a particular role under aggressive pressure from a superior member (e.g., herding).

8. Individual traits and affinities: The blend of individual traits in a group and individual friendships and aversions can shape the group within the range of species variability.

These are the variables or dimensions of group-structuring behavior. Species and populations have varying intensities of each behavioral set and differ in the sex and age classes which they subject to each of these tendencies. The combination of intensities and selections thus determines the specific type of their society.

The above list is only tentative. Our ultimate aim is a thorough understanding of the inherited behavioral potential of primate species and genera. With this understanding we shall be able more realistically to evaluate the successes and failures of a species in meeting the challenge of its environment than if we simply postulate that each observed trait is adaptive. We may then understand why an adaptation may be less than ideal, why it is the best solution of which the species was capable, and how much the species has to pay even for this next-best design. A way of life will then appear as a compromise, first between the conflicting requirements of the environment, and then between these requirements and the limitations of the animal's own behavioral resources as they emerged from its phylogenetic past.

ADAPTATION BY TRADITION

Japanese Macaques: Feeding Traditions on Koshima Island

The little island of Koshima is a wooded, precipitous mountain surrounded by sandy beaches and the sea. Until recently only the mountain and its forest had any ecological significance for the group of Japanese macaques (Macaca fuscata) inhabiting the island; they had so far not foraged on the beach and they had never entered the water. In 1952, however, the researchers of the Japan Monkey Center began to feed the troop on the beach of the island and thus triggered an ecological expansion that provided some fascinating insights into the adaptive potential of primates. The following description is based on a detailed report by Kawai (1965).

The artificial feeding consisted of throwing sweet potatoes onto the beach. The group soon got used to leaving the forest and to eating potatoes as free of adhering sand as possible (Fig. 4.7). The beach became not only a new foraging ground, but also the breeding ground of what the Japanese researchers call a "preculture." One year after the feeding was started, a nearly two-year-old female named Imo was observed carrying a sweet potato to the edge of a brook. With one hand she dipped the potato into the water while she brushed off the sand with the other. In the years to follow, the technique slowly spread throughout the group. In addition, the washing was gradually transferred from the brook to the sea (Fig. 4.8). Today potato-washing in salt water is an established tradition which infants learn from their mothers as a natural adjunct of eating potatoes.

The new habit was transmitted from monkey to monkey in two distinct forms. "Individual propagation" first transmitted the habit from the young to the adults of the existing generation. The ease or resistance with which a monkey took to potato-washing in this phase was partly a function of sex and age. The class to learn most readily was that of the juveniles between one and two-and-a-half years, Imo's own age class. Male and female juveniles learned with equal readiness. Five years after Imo started the behavior, nearly 80 per cent of the younger group members in the age class of two to seven years washed their potatoes. The adults above the age of seven were more conservative. Only 18 per cent of them had acquired the new behavior and all of these were females. The remaining adults never adopted the habit.

These results, however, could not be explained by adult and male conservatism alone. A close social relationship with a performer was also essential. Potato-washing was not acquired by observing some distant group member, but only when feeding with an intimate companion. Thus, mothers

Fig. 4.7. The Koshima group of Japanese macaques foraging on the beach on the island. (Photograph by M. Iwamoto.)

Fig. 4.8. Potato-washing in the sea is one of the newly invented behaviors of the Koshima group. (Photograph by M. Kawai.)

readily learned from their children and older siblings from younger brothers and sisters. Subadult and young males, on the other hand, had few opportunities to feed side by side with potato-washers; they stay mostly on the periphery of the group, whereas the youngsters and their mothers live in the central part.

By the time a male returns to the center as a leader, he is apparently too old and inflexible to change his ways. But lacking the opportunity to feed with performers is hardly the only cause of male resistance to the new behavior. One year after Kawai's report, Menzel (1966) published a study on the response of feral (wild) Japanese monkeys to strange objects placed on one of their trails. Juveniles responded much more frequently to a yellow plastic rope, for example, than did adults. Up to the age of three years, males reacted as often as females, but among adults, the females responded in 48 per cent of the cases, adult males only in 19 per cent. A typical "nonresponse" of adult males was an almost imperceptible angling off from the line of travel and a mere glance sideways toward the rope. Obviously, detectable responses to the yellow rope and readiness to adopt the new habit of potato-washing are similarily distributed among the sex and age classes, although the first is an individual action that does not require a close social affiliation with a model. Some as yet unknown factor in the behavioral setup of the adult male seems to suppress responses to novel stimuli regardless of their social context. Among Japanese macaques, at least, the adult male shows little disposition to be an active promoter of new behavioral adaptations. The apparent acceptance of such a role by the human male may be a result of the neoteny of our species, that is, of the tendency to preserve juvenile traits of behavior into adulthood.

At Koshima, potato-washing mothers passed the behavior to all their infants born after they had themselves acquired the habit and thus initiated the second phase, which Kawai calls the "precultural propagation." At this stage, expansion reversed its direction and began to pass from the old to the young. In addition, the behavior was now acquired in a new way. Whereas juveniles and adults had so far picked up the complete washing behavior at once, the infants now learned bit by bit. Earlier generations had never entered the water, but the new babies were now taken into water clinging to their mothers' bellies. By the time they began to eat solid food, they were fully habituated to the medium that their ancestors had avoided. At the age of six months, they began to pick up from the water pieces of potato which their mothers had dropped. The complete washing behavior developed by one to two-and-a-half years, also the age that had proved most accessible to the new behavior at the stage of individual propagation.

The second way of acquiring the behavior had an interesting secondary effect: All potatoes that the new infants ate were seasoned with salt water, and the taste of salt apparently became associated with potatoes. Many of the new generation now not only washed their potatoes, but also seasoned them by dipping them into the sea between bites.

The Koshima group had still more in store. When the scientists began to scatter wheat on the beach, the female Imo, now four years old, invented another trick. Instead of picking the grains singly out of the sand, she carried handfuls of mixed sand and wheat to the shore, threw the whole mess into the water, and waited for the sand to sink and the wheat to float; then she collected the wheat and ate it. Again, the habit spread among juveniles and their mothers, and again the adult males would have nothing to do with it. There was, however, one interesting difference: Potato-washing had been most readily acquired by one to two-and-a-half-year-old monkeys. In contrast, most of the monkeys who acquired the wheat trick were of the age class between two and four years. Note that Imo herself was one-and-a-half years when she invented potato-washing and four years when she introduced wheat-washing.

Perhaps, then, new behaviors are most readily copied from peers. There is, however, another possible explanation that must remain speculative but that raises significant implications. The main component of potato-washing is brushing the tuber with one hand or, in another form, rolling it between the hand and the ground. These are the same movements baboons use when removing dirt or bristles from a fruit. Apparently these behavior patterns develop easily from the behavioral potential of ground-living primates, or, in other terms, these species are genetically predisposed for brushing and rolling movements. In this case, potato-washing is new only insofar as the cleaning movements are performed in water. Wheat-washing, instead, contains an element that is definitely not an easily realized part of the primate's behavioral potential; food already collected must be first thrown away before it is eaten. This would come more easily if primates were used to hoarding food and thus to abandoning it temporarily, or to sharing it, or at least to carrying it away before eating it. Wheat-washing behavior may thus depend on a higher degree of behavioral maturity and therefore be acquired later in life than potato-washing.

Whether or not these explanations are correct for the particular case is not my concern here; what I wish to exemplify is the type of the explanation: that the possibility of a behavioral innovation depends in part on the genetically determined range of potential modifications, and that different behavioral modifications are related to different stages of individual maturation in which they are realized most easily. Kawai himself rightly refrains from attempting an explanation.

We have seen that newly acquired behavior is most readily transmitted among animals with a close social affinity. The pathways of habit propagation follow, as it were, a preestablished network of affinities within the group and reveal a structure of subgroups with frequent positive interaction. In the Koshima group, habit propagation was greatly influenced by kinship. Entire "lineages" consisting of a mother and her descendants tended to acquire or reject a new behavior as a unit. Between 1951 and 1960, for example, the sons and daughters of the female Eba acquired an average of 3.6 of the various new behaviors invented by the group in this period, whereas Nami and her descendants acquired only 1.6 new habits per individual. The children tend to be as receptive as their mother, but it is as yet unknown how much of their similarity stems from their common genes and how much is due to their learning from the same mother.

Among the most interesting aspects of the Koshima events are the secondary effects of the new traditions. The changes in feeding behavior reverberated into superficially remote parts of the socioecological system. The habit of washing primarily facilitated the rapid ingestion of food, but beyond this it opened the way to a hitherto irrelevant part of the habitat, the sea. The youngsters of the new generations took up bathing as part of their playful and exploratory activities. Splashing became a preferred pastime in hot weather. The juveniles learned to swim; some of them began to dive and brought up seaweed from the bottom; at least one of them left Koshima and swam to a neighboring island. The sea thus became a potential food source, and it was no longer an absolute barrier to would-be migrators or to socially hard-pressed refugees from the island group. One habit had prepared the way for expansive changes in the group's ecology and social structure. The stable system has now entered a phase of changes which perhaps will trigger still more changes, until the possibilities of the habitat and their behavioral limitations interlock in a new stability. New selective pressures could begin to affect the gene pool and induce further changes, such as extensive foraging in the sea.

The Koshima macaques may be so limited in their behavioral potential that a new stability will soon be reached. But it can be imagined that a primate species of quite another kind might never reach stability again: Changes in its habitat-oriented behavior might breed even more changes that would finally increase exponentially into a runaway development. This, it seems, is happening to our own species.

The Role of Tradition

The propagation of newly invented forms of behavior as exemplified by the Koshima macaques is the closest primate parallel of human culture studied in detail to date, although it is a tradition of behavior forms only, not of symbols.

In Chapter 1 we introduced the distinction between genetic and acquired differences of behavior, that is, between phylogenetic adaptations and adaptive modifications. We must now recall how we subdivided the second category. A form of behavior can be acquired by the individual interacting with and learning directly from the nonsocial environment. Imo learned the washing behavior by interacting with food, sand, and water; fishing technique can be acquired simply by observing fish and by trying to catch them; this we called an ecological modification. Or one can acquire behavior by social modification, that is, with some help from conspecifics, either by directly imitating them, or by being socially rewarded for a particular kind of spontaneous behavior, or by having one's attention focussed by the activities of others on a certain part of the environment—a mechanism called "local enhancement" by ethologists. (It is beyond the frame of this text to discuss the mechanisms that induce an animal to behave like his partners. The interested reader is refered to the textbook by Thorpe (1969). Here, it should only be understood that imitation is not the only possibility.) Depending on the mechanism, the individual will in the end copy either the detailed movements of his conspecifics, or their tendency to do certain things while refraining from doing others, or only the direction of their interest.

Not every social modification of behavior is also a tradition. If "tradition" were used in this widest meaning it would include a male macaque's technique of copulation, which is known to develop only in a normal social environment. The term tradition may reasonably be limited to behavioral modifications for which other populations have developed viable alternatives.

Tradition learning on Koshima has not been fully analyzed, but from Kawai's description it seems that imitation was assisted by several other mechanisms. Our question in this section, however, is not how exactly tradition perpetuates culture, but what its specific adaptive advantages are. Under what conditions are adaptations based on tradition more efficient than the other two kinds of adaptation, namely individual learning and genetic change? Deduction leads to the following answers:

1. Tradition is superior to individual learning if the new behavior is difficult to acquire individually in direct interaction with the environment. Not every macaque is so inventive as Imo; in addition, different members of the society are endowed for different types of individual learning. Tradition can pool their individual achievements.

2. Experimenting directly with the environment may be dangerous, as with poisonous food plants or predators. In these cases, tradition is the safe way of acquiring adaptive behavior. In fact, young primates feed only on what they see others eat. Wild hamadryas baboons who have never seen bananas do not dare touch them at first, although they eat them eagerly once they have been introduced to them in captivity. In many species, juveniles sniff the mouth of older group members who are chewing food, probably to inform themselves what their elders are eating.

3. Some environmental situations such as drought are too rare to permit a direct experience for every troop member. In this case, an experienced oldster may be the only animal in the troop that has the adaptive response—which is to travel to a certain lasting waterhole outside the usual home range and thus inform the younger members of its whereabouts. Tradition here requires a long life expectancy and a leading role for the older animals. It is also advantageous if a single experience is sufficient for establishing the adaptive behavior.

Difficult, dangerous and rare individual adaptations are favorably perpetuated by tradition. Thus an experience is multiplied without the high costs of individual adaptation. Exactly the same advantage, however, could also be obtained if the adaptive behavior became genetically programmed, and we must now try to outline the conditions where tradition is superior to phylogenetic adaptation.

The obvious answer is that mutation and selection often take too long to build up new behavioral adaptations. In addition, genetic programming of a trait can occur and is only adaptive only if the corresponding condition of the environment persists for many generations and is the same throughout the habitats of the entire population in which genes are exchanged. Adaptations to local conditions that persist for only a few generations are more aptly maintained by tradition. If these behaviors are at the same time difficult to acquire by direct individual learning, then a tradition of individually acquired behavior must be the method of choice, the most adaptive among the adaptive mechanisms.

It is easy to enumerate behaviors that probably fall into this middle class between the domains of individual learning and evolution. The local topography of sleeping-sites, water-holes, and food sources, especially in years of extreme weather conditions, the species of desirable or poisonous plants and animals, and the lairs and habits of local predators cannot be learned by every troop member without great risk and delay; they also vary too much in time and from one home range to the next to permit genetic coding of the relevant information to occur or to be useful. Tradition is thus the appropriate vessel for information that is relevant only to a few generations within a limited area. In contrast, direct individual experience with the environment is appropriate if the information is relevant only to the individual who has the experience and if this information is easy to discover. An example is the knowledge and appropriate treatment of group members and of one's social position in the structures and hierarchies of the group. Genetic coding, consequently, is reserved for behaviors that prove adaptive throughout the entire population or species over millennia, such as clinging to one's mother or seeking out a place above ground for the night.

Although tradition functions through individual learning, it shares an interesting aspect with the process of phylogenetic adaptation. Both processes spread information through a population. The question that arises is, how fast, optimally, should the old information be replaced by the new? The general problem can be exemplified by a troop of baboons that one day finds for the first time a number of baited traps on its daily tour. The bait is soon discovered, but the strange structures containing it are strongly avoided. A few juveniles finally enter the traps and begin to feed, while the rest of the troop watches. How fast should the apparently adaptive new behavior of eating food in a wire cube spread? A new behavior can obviously spread too rapidly, that is before its adaptiveness is fully tested. Similarily, on the level of phylogenetic adaptation, geneticists warn against an increase of mutation rates by artificial radiation, although such an increase would raise the number of adaptive as well as maladaptive mutations. In both examples, the adaptive value of a particular item of new information is unknown, and, in comparison with the well-tested pattern that would have to be replaced by it, highly questionable. New information is therefore treated as a risky experiment to which only a small fraction of the population is subjected. Juveniles are the obvious candidates for the experimental fraction since they are the most easily replaced investment of food and experience.

The optimal size of this experimental fraction depends on how predictable the effect of the innovation is. In genetic mutation, the probability that an innovation is harmful is almost 1.0. Thus only small mutation rates are tolerable. When a corporation introduces a new accounting system, the insight of the planning experts greatly reduces the probability of its failure, and the executives may decide to subject perhaps one fourth of the company's plants to the experiment. On Koshima, only 4 of the 60 macaques acquired potato-washing in the first year after its invention. We can imagine that risky new behaviors spread more slowly through a troop than objectively safer types of behavior; we do not know. But there is little doubt that conservatism, too, is adaptive. The inflexible adults of the Koshima troop form a safety reservoir of the previous behavioral variant, which will survive the invention for at least ten years. If the new behavior should turn out to be harmful, say because of parasitic infection, they would survive. In spreading new behaviors, adult rigidity has the same function as low mutation rates in evolution.

SUMMARY

1. The adaptive potential of a species is limited by phylogenetic dispositions. New behavioral adaptations are possible only if the necessary dispositions are within the scope of the behavioral heritage, and if they can be accommodated within the existing social system. Distinctions between ideal and merely tolerable adaptations require causal research on how adaptive traits are constructed.

2. In evolving their present social system, hamadryas baboons introduced two new social units, the troop and the one-male group. Troop life, which requires no more than increased tolerance among groups, is an easy modification found in other baboon species as well. The hamadryas one-male group is based on male herding behavior. Its behavioral components were already present in the baboon genus, but their organization into the herding syndrome required phylogenetic adaptations. The female's following-response, in contrast, can be experimentally identified as a modification. The major secondary adaptation within the new social system is an inhibition of males to encroach on each others' females. The hamadryas system is compared with hypotheses on the origin of human pair bonds.

3. Other species build their one-male groups with different behaviors. Among geladas, the dominant females assist the male in preventing intergroup contacts. Patas one-male groups, living apart from each other, have no such problems; their one-male groups separate because the males do not tolerate each others' proximity. Geladas and hamadryas form troops because their males are mutually attractive and capable of submission. The presence of females counteracts attraction among males.

4. Japanese macques offer examples of a special case of modification: Tradition. Potato- and wheat-washing were invented by one individual and imitated by the group. Adults were more resistant to the new behavior than were juveniles, and adult males more than adult females. Close social ties facilitated the transfer of the habit. The new tradition had secondary effects: Youngsters raised in the tradition also acquired swimming, diving, and the seasoning of food in salt water.

5. Adaptation by tradition is more adaptive than individual learning when acquiring the new behavior is difficult, or dangerous, or rarely possible. Tradition is superior to phylogenetic adaptation when behavior has to adapt to rapidly changing or merely local conditions.

Chapter 5

HOW FLEXIBLE IS THE TRAIT?

Some of the most fascinating opportunities for the study of behavioral adaptations in nature occur where two species with common ancestry but with different organizations come into contact. When we learned, at the beginning of our second field study in 1968, that the newly founded Awash National Park in Ethiopia was inhabited by both hamadryas and anubis baboons, we immediately decided to work there. In the following days, the discussions in our team focused on the possible nature of the species border between the family-forming hamadryas and their family-less relatives. Why was the border at Awash instead of somewhere else? Would an analysis of the border zone give us some hints on the ecological factors that favor the specialized organization of the hamadryas? If there were an ecological advantage in the one-male group system, the hamadryas would probably expand just as far westward as the habitat conditions supporting it. In the absence of physical barriers, anubis baboons would have to take over exactly where the changing environmental factors began to favor a non-family organization with large, closed groups of the anubis type. Before we took up the project, we developed a simple model in which two opposed gradients of success met in a zone of equal survival value for both species.

At Awash, we found that there is indeed an ecological transition. Upriver and westward the Awash River flows level with the surrounding flat thornbush vegetation. On its borders, the river supports a gallery forest with trees up to 60 feet high. This whole area is inhabited by anubis groups that roost in the gallery forest every night but make daily foraging trips into the thornbush. This picture changes abruptly at Camp Awash. The huge Awash Falls send the river into a canyon that steadily deepens and widens on its eastward course toward the Danakil desert. The steep slopes of the canyon support only a meager, interrupted gallery forest, overlooked by the endless cliffs of the canyon (Fig. 5.1). In the canyon, we first encountered only hamadryas baboons who spent the night in cliffs. According to reports from other parts of Africa, anubis baboons sleep mostly in trees, and we had so far never seen hamadryas sleep anywhere but in cliffs. The answer to our question looked disappointingly simple: Trees but no cliffs meant anubis, cliffs but no trees meant hamadryas.

Our team was later joined by Ueli Nagel, a graduate student who had decided to give the species border special attention. His quantitative comparison confirmed that the major difference between the habitats above and below the falls lay in the proportion of trees and cliffs. The gallery forest above the falls covered a higher area per unit length of the river than below, and the number of tree species in the forest declined from 14 species above the falls to 11 below the falls and to 8 farther downriver in the canyon. However, the ecological transition was noticeable only along the river; the inland thornbush and grasslands, where all baboons collected part of their diet, did not significantly change from west to east. Visibility in the bush on the level of a standing baboon was poor in both sections, averaging about 10 yards.

Nevertheless, Nagel did not take long to shake our hypothesis: The first baboon group immediately below the falls was

Fig. 5.1. The Awash Canyon below the Falls, with cliffs and thin gallery forest. The slopes and the plateau are covered with acacia thornbush. (Photograph by U. Nagel.)

an anubis group and it slept in cliffs, although the local gallery forest offered some apparently good roosting-trees. Its
social organization, however, gave no evidence of one-male
groups; it was typically anubis.

Why was the first group below the falls an anubis group,
and why was there no other farther down? Nagel argued
that the group's presence in the canyon was quite compatible
with the assumption that anubis baboons fare worse under
canyon conditions than hamadryas. If the anubis above the
falls were more successful and reproductive than the hamadryas baboons in the canyon, then excess anubis baboons
might spill over the ecological barrier and support a group
under poor conditions by constant immigration. In fact, this
group contrasted with all others by its high incidence of
skin disease. However, the spillover of anubis genes across
the ecological borderline was telling us less about what was
adaptive than about the flexibility of adaptive traits. In their
sleeping habits, the cliff-roosting anubis adapted to local conditions, but in their social structure they remained true to
their species, that is to their genotypes or, perhaps, to some
rigid tradition.

Nagel's study soon brought a second surprise. The three
next groups downriver from the canyon anubis were not
hamadryas, but hamadryas-anubis hybrids of all shades. The
first pure hamadryas lived 15 miles below the falls. All the
hybrid groups roosted on cliffs as did the hamadryas farther
down, but in their social organization they again seemed to
follow their genes, not the habitat. The social organization
of the hybrids revealed a curious mixture of anubis and
hamadryas characteristics. There were one-male groups, but
most of them were small and unstable. Although some males
were as active herders as pure hamadryas, their success in
forming harems was meager; perhaps their threats and attacks on females were imprecisely correlated with female
following or straying.

Thus the ecological transition from forest to cliff country

corresponded exactly with a change of baboon sleeping habits, but social organization was not directly affected by the ecological change. Anubis groups without one-male groups, hamadryas troops with one-male groups, and the in-between societies of the hybrids all occurred in the same canyon habitat. This suggested that either species of baboons was capable of modifying its sleeping habits to local conditions but maintained a relatively inflexible social organization.

Since changing a social system is certainly a more complex process than taking a cliff for a tree, this difference in genetic involvement is not surprising. A number of observations support this impression. First, no anubis population in Ethiopia or in other parts of Africa has yet been reported to form stable one-male groups, and we have yet to find a hamadryas troop without one-male groups. On the other hand, cliff-roosting anubis are known in several African areas where high trees are scarce. On the Serengeti plains, for example, a group of anubis have colonized a rocky outcrop within a vast, nearly treeless plain. A group of anubis baboons in Ethiopia even roosts occasionally on iron powerline masts. As if to complete the evidence, we found a hamadryas troop in the flat, cliffless Cassam valley north of the Awash that spends the night in high trees along the river. Even so, the Cassam hamadryas are organized in one-male groups. The border situation at Awash thus provides an example of the two ways of adapting outlined earlier. Roosting habits demonstrate a case of modification, while the social structure of the two species suggests the workings of phylogenetic adaptation.

The fact that the species border did not coincide exactly with the ecological transition allowed a further sorting-out of species-typical and habitat-typical behaviors. The length of the daily route was found to belong to the latter class; there was no significant difference between the average 3.3 miles of the anubis above the falls, the 4.3 miles of the canyon hybrids, and the 4.0 miles of the hamadryas farther downriver, but these figures differed significantly from the 8.2 miles of the hamadryas population living further east in the arid Danakil plain.

Interesting differences occurred in the preference of the groups for the forest parts of their respective habitats. Hamadryas spent significantly less time in the gallery forest than did the anubis above the falls, presumably because the forests offer a smaller proportion of the food in the canyon habitat. The hybrids spent as little time as the hamadryas foraging in the forest; as to resting there, they took an intermediate position between the pure species. Apparently their feeding activity was adapted to the canyon habitat which they shared with the hamadryas. But resting, which can be done anywhere, seemed to reflect inherited preferences for forest or open country (Fig. 5.2).

The splitting of troops at roosting time is another example of plasticity. The anubis baboons in the gallery forest above the falls slept in united groups. In contrast, all baboon troops in the canyon, including the anubis, roosted in separate parties of unstable numbers. This species-independent trait suggests a local modification. A large group that can split into parties of varying size seems adaptive in a habitat of the canyon type, with its limited and scattered food supply and its many sleeping-sites. Judging from inland populations of either species, it so far appeared that stable sleeping groups are typical of anubis baboons while varying roosting parties are a hamadryas trait. At the species border, it became clear that anubis baboons can also split and reassemble in various combinations, although their social structure seems less suitable for the process.

Similar insights can be obtained from areas where two related species share the same habitat, where, in zoological terminology, they are sympatric. An example is the study by the British researchers John Crook and P. Aldrich-Blake (1968) on a relict forest habitat at Debre Libanos in the Ethiopian highlands. Papio anubis here appears once more in a marginal habitat, but in this case on its border toward

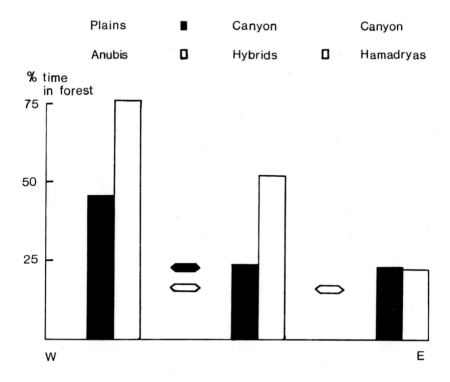

Fig. 5.2. Percentage of feeding time (black bars) and resting time (white bars) spent in forests by anubis baboons, hybrids, and hamadryas baboons in a section across the species border. ■ marks the transition from the plains habitat to the canyon habitat; □ indicates a transition of genotypes. Double arrows (◆▶◁▷) indicate a significant (0.01) difference of percentage. It is evident that the frequency of feeding in the forest varies in correlation with the habitat, whereas the frequency of resting in the forest varies with the genotype. (From Nagel, in press.)

the open grassy mountains, not toward the lowland semi-deserts as in the Awash region. Its neighbor here is not Papio hamadryas but Theropithecus gelada. Like hamadryas, geladas occupy extreme habitats with few or no trees. Both species roost on cliffs and both are organized into one-male groups that congregate into large troops. The anubis with its multi-male groups occupies the rich, originally forested mid-

dle section between the cold highlands and the arid low-
lands.

Unlike the two Papios, the gelada and the anubis do not
interbreed in the wild. Therefore, the two pure species can
jointly occupy the same habitat, and they do so in the study
area around the monastery of Debre Libanos, situated on the
slope of a deep gorge of the Blue Nile system, 43 miles north
of Addis Ababa. Ecologically, it includes three sections: (a)
a small forest, preserved as church land, and a relict of the
original vegetation before its degradation by man; (b) low
scrub loosely covering the steeper slopes, interspersed with
vertical cliffs; and c) agricultural land and sward, both of
which turn sterile and dusty in the dry season.

The three sections were jointly inhabited by about 80
anubis baboons and about 300 geladas, but the two species
clearly differed in their preferences. Nearly half the anubis
parties of a random sample were sighted in the forest; only
a few were found in the open scrubland and fields. The geladas, in contrast, stayed mostly in the open and avoided the
forest. This was not because one species aggressively dis-
placed the other from certain parts of the habitat. Anubis
and geladas were not observed fighting or chasing each
other; for periods of several minutes they even formed mixed
parties. Their preferences, therefore, are effects of choice
based on species-specific tendencies.

Associated with the preference for a particular vegetation
type were other behavioral differences that revealed ecological significance. The geladas never climbed a tree, even
when they happened to stay in the forest. Anubis baboons
frequently rested and played in the forest trees and plucked
their fruits. When caught by heavy rain or hail, the geladas
dashed for an overhanging cliff, while anubis climbed trees
with a thick canopy. In their diet, geladas chose the bulbs,
roots, seeds, and leaves of the open grasslands; even when
near trees, they preferred the small, dried olive seeds lying
on the ground to the complete olive fruits, palm fruits,
prickly pears, and fresh twigs eaten by the anubis baboons.

The gelada fed by digging and collecting small items while sitting and shufflling forward on their haunches; the anubis strolled, selected a fleshy piece, and sat down to divide and chew it. The typical skill of gelada feeding was to collect rapidly a neat bundle of grass leaves with one hand, whereas the anubis baboons were experts at removing the spines and prickles of opuntia leaves and fruits by plucking and rolling.

The dentition of geladas is specialized for grinding small hard items; this definitely "innate" trait suggests that their food preference is somewhat narrow, causing them to disregard easier food even where it is available, as at Debre Libanos. Crook and Aldrich-Blake could actually demonstrate that gelada food requires more effort to assure proper nutrition: At least between 10 A.M. and 5 P.M., geladas spent 35 to 70 per cent of their time in feeding, whereas the anubis so spent only 20 per cent. Finally, the preferred vegetation types were related to the escape behaviors of the two species. When disturbed by humans, the geladas uttered piercing alarm cries and bolted for the cliffs. Being clearly visible on the open fields, they had nothing to gain from furtiveness. In contrast, once a discovered person approached them, the anubis stopped barking and silently slunk away from the bushes and trees.

These differences indicated the parts of eco-behavior that are relatively fixed species traits. But the Debre Libanos study also revealed life patterns in which the two populations behaved alike but deviated from the normal behavior of their species elsewhere. These traits displayed the flexible parts of their adaptations, that is, the range of their modifications. For example, the anubis again slept in cliffs as they did on the hamadryas border at Awash, Furthermore, both geladas and anubis formed small social units of varying sizes during the day. In an average anubis habitat, groups comprise about 70 individuals, while gelada troops in their high mountain range in Semyen average more than 100 animals; at Debre Libanos, both species were usually encountered in small groups of 15 to 20 members. Typical anubis groups maintain constant membership for at least a few days or months; at Debre Libanos, there was an almost constant re-shuffling.

Both species seemed to live under suboptimal conditions in a marginal habitat; again, as in the Awash example, each species was able to modify its group size but not its basic social system. The anubis adhered to the promiscuous system, although their usual unified groups were broken up. Their foraging parties were quite unstable. The geladas, on the other hand, adhered to the one-male system that Crook had earlier found among their central populations in the Semyen mountains. Most of the small gelada parties at Debre Libanos were one-male groups; they never numbered fewer than four members and were stable over time. In discussing the border situation at Awash, we hypothesized that a system of one-male groups is advantageous because its provides pre-formed division lines. The Debre Libanos case suggests fur-ther advantages: The one-male groups are stable minimal parties; their stong cohesion sets a lower limit to splitting; and their composition assures that each party includes at least one large adult male.

Species borders of the Awash and Debre Libanos type are interesting because they show genetically different popula-tions in their adaptation to the same or similar habitats. In the two cases, we have seen that certain behavior traits, such as cliff-roosting, vary with the habitat, but independently of the species concerned (Table 5.1). In these local modifica-tions, the border populations resemble each other and differ from their conspecifics in more typical areas farther inland. In general, modifications can be expected to reflect local con-ditions.

Other behavioral systems, notably the social organization of the three species, do not vary with environmental condi-tions but are strictly species-typical. They occur even if the neighboring or sympatric species maintains a different social system that seems more adaptive under the local conditions. This rigidity, though possibly owing to tradition, is probably

Table 5.1. Tentative distinction between species–typical and habitat–typical behavior of three baboon-type species, as suggested by their behavior at common species borders.

| | Behaviors correlated with | |
	local habitat but not with species	species but not with local habitat
Anubis- hamadryas border (Awash)	group size and stability foraging time in forest length of daily route type of roosting place	social system resting time in forest
Anubis- gelada border (Debre Libanos)	group size and stability type of roosting place	social system preference for dense or open vegetation feeding technique and food selection escape behavior
Interpretation	relatively flexible traits with loose genetic control: modifications, permitting adaptation to local con- ditions	relatively rigid traits with close genetic control: phylogenetic adaptations, permitting adaptation only to average condi- tions in total species area

an effect of genetic control, that is, of phylogenetic adaptation. Baboons may simply be incapable of producing their social systems without detailed assistance from the genetic code. The price of this assistance is that they are unable to change rapidly between two or more social systems. Although adapted to the average environment of the species, phylogenetic "adaptations" can be definitely maladaptive locally. The population will nevertheless survive if its genetic package is on the whole successful. In such cases, the often made assumption that a particular trait is an adaptation to a particular habitat simply because it occurs there is obviously incorrect.

Adaptation research has two distinct tasks. The first is to show the adaptive function of a trait in each of several habitats. The second is to determine how widely each trait can vary on the basis of the existing heritage and thus to delineate the population's adaptive potential. Marked geographical transition zones of species or habitats can help with this second task.

SUMMARY

Where two species meet on a common border, the transition between the two genotypes is often more abrupt than the transition between the two habitats. In some of their behaviors, the border populations of either species may then converge on a similar way of life, obviously by modifying these behaviors in adaptation to the border conditions. In other behaviors they do not adapt to the local conditions but behave just as their species behaves everywhere in its range. The transition across the border then parallels the abrupt change of the genotype. Thus a species border often permits a preliminary distinction between modifications and engrained, phylogenetic adaptations. The social systems of baboons apparently belong to the second class.

Chapter 6

MAN AND PRIMATES COMPARED

DISTRIBUTION

There are many signs that indicate the success of a species; one of them is simply the size of its geographical range. Even a superficial look at what primates and man have achieved in inhabiting the earth shows that the range of Homo sapiens includes the ranges of all other primate species taken together. It actually extends far beyond their total range, especially toward the poles but also away from the continents to the oceanic islands.

Distributions further suggest that man in some way stands apart from the rest in his ecological needs. In the African rain forests many primate species are sympatric. They can survive in the same habitat because it offers a variety of ecological niches, and so it is not surprising that there is also a niche for man. The picture changes as we leave the rain forests and move into the savannas; the number of primate species that share a common habitat now declines to two or three. Man is still with them. Finally, in the semi-deserts and the colder areas, we find only one monkey species in each area. The semi-deserts of the Red Sea coasts are inhabited only by hamadryas baboons, the cold and barren Ethiopian mountains only by geladas; only anubis baboons are found in the desert mountains of Tibesti, and only the Barbary macaque in the Atlas range. Since no physical barrier prevents other species from colonizing the hamadryas and the gelada areas, we must conclude that their habitats offer primates only one ecological niche. But even here, we find man. Apparently his ecological niche is so different from that of his pioneering fellow primates that both may exist in the same harsh environment.

Traveling farther north, far beyond the ranges of the last nonhuman primate, we continue meeting that same species man that we first saw among scores of other primate species in the equatorial rain forest. Ever changing his survival technique, he seems to fit into every niche on dry land. Nearly naked in the rain forest, man becomes hairy with the hair of animals in the north, and in the artics, wooden extensions on his feet carry him over the snow and make him, ecologically, a new animal. The tools with which he transforms himself and his niches are his cultures. The geographical mosaic of the nonhuman primates, confined in their narrow species ranges, restricted to their single way of life, contrasts with the pervasive distribution of ecologically polymorphic man.

DIVERSITY OF SOCIAL STRUCTURE

Since ecological polymorphism requires polymorphism of behavior, wide adaptive success should also reveal itself in a variety of social structures. Because the human species survives in more habitats than all other primate species together, one would expect that its social behavior, as far as it is related to ecological conditions, would vary accordingly. Man would then equal the behavioral range not of one but of many primate species. This seems correct in at least one respect: One primate species generally adheres to a single type of social structure. For example, all gibbon populations studied to date are organized into monogamous pairs of one

male and one female plus young. So far, all known hamadryas baboons and geladas live in breeding units of one male
and several females, and savanna baboons are always organized in larger, promiscuous units. In contrast, the social
structures of our own species range from monogamous to
polygynous systems and even include polyandry, which is
unknown among primates.

Nonhuman primate species may, however, not be quite so
homogeneous in structure as would now seem, since only a
few populations of each species have been studied. A larger
sample might reveal the odd aberration which, like polyandry in man, would not be discovered in a brief survey. We
know already that Indian langurs (Presbytis entellus),
which form multi-male groups in many areas, can also orga-
nize into one-male groups in certain other regions.

If the present impression that most primate species have
only one social organization is confirmed, it might be taken
as proof of a narrow modification range. But such a conclusion would be premature. Man's technological success has
exposed his social behavior to a far greater variety of environments than any other primate deals with. Such variety
must have, at one or another time or place, activated nearly
every behavioral modification of which humans are capable.
If other primate species were exposed to a similar variety of
modifying influences they might also reveal a broad potential
for diverse social organization. On the other hand, the evidence of hamadryas baboons suggests that the genetic potential of some nonhuman species may indeed be restricted
to one type of society which environmental change would
hardly alter.

TECHNOLOGY

Man is the animal that not only occupies but also
shapes its ecological niches by means of technology. When
inspecting the known primate achievements in this respect,
one can only be unimpressed. Their lack of elaborate techni-

cal skills compares unfavorably with those of many so-called lower vertebrates and many invertebrates. Hundreds of bird species build nests a hundred times more elaborate than the chimp nest, which is the highest achievement of primate building activity. Certain weaver birds build roofs above their nest colonies; primates at best use one when it is already there.

The comparison is somewhat unfair, because the evolutionary trend from birds to mammals tends to replace parental behavior by parental physiology. Primates, being mammals, raise their young first within the mother's body and then feed them with her milk, which makes complex nesting and feeding behavior unnecessary. But rodents are mammals too. Many of them dig burrows and pat them with plant parts carried into the nest from outdoors. Many collect and hoard food. There is not one among the two hundred or so primate species that constructs an ever-so-simple burrow or does much with food but eat it on the spot. Specialized swimmers and divers have evolved among both rodents and carnivores, but not among primates.

Primates are, superficially, as unspecialized and primitive as their insectivorous, bush-dwelling ancestors. They occasionally catch prey to eat, but their hunting techniques cannot match those of the specialized carnivores. They are primarily vegetarians; but they can neither survive a severe winter at high altitudes or latitudes, nor can they go long without water in dry areas, as some ungulates do. Although primates have prehensile hands, their use of tools is modest. Chimpanzees poke at hidden insects with long, thin stalks, but so does a species of Galapagos finch. Some macaques are reported to smash shells with stones. Egyptian vultures open ostrich eggs the same way, and diving California sea otters carry a stone with every mussel to the surface of the sea and there crack the shell. Nobody would have predicted that a primate would develop a technology of human dimension.

Given that no nonhuman primates hunt, build, or store food as elegantly as, say, wasps, how do we interpret the fact that it was a primate that ultimately developed all these behaviors to their greatest extent? If monkeys and apes are at all more human than other mammals with regard to their behavioral substrates, then why have primate field studies depicted their subjects as such poor performers in these domains? And what pressures would have favored the evolution of mental capacities that apparently are not used?

One answer is that some of the most developed primate abilities are manifested only in critical situations so rare that they have never been witnessed by field scientists. Some of the fantastic hunter's tales of baboon burials and other unusual behaviors may contain a grain of truth. Field observation is indeed a poor method for studying rare but important behavior. It takes experiments creating unusual situations to delineate a behavioral repertoire that comes close to the animal's full capacities. No field observation, it seems, could ever have predicted that chimpanzees would collect metal coins of certain colors if they could later exchange them for fruits in a slot machine, but laboratory experiments have shown that they have this ability. A vast body of experimental data has demonstrated an unusual disposition in chimpanzees for tool-using (Fig. 6.1), technical problem solving, and cooperation, essential capacities in human technology.

In contrast to the highly specialized but rigid skills of lower vertebrates, then, primates have a potential for learning broad sets of tasks which neither they nor their ancestors encountered in this particular form. *This flexibility, and not a specialized but genetically fixed skill, prepared the way for culture.* Thus, monkeys and apes are not so far removed from man in their capacities as the study of their everyday life in the wild suggests. Even so, the primates' success in laboratories does not explain why they ever evolved abilities that seem so unimportant or even inapplicable in their habitats.

Among the tentative answers to this problem, I shall report the one I find most convincing. The British ethologist Michael Chance has repeatedly discussed the hypothesis that

the large primate cortex and the corresponding ability to use
new tools may first have evolved in the context of social behavior and not in the context of technical exploitation of the
habitat. Most primates are sexually stimulated and motivated
for many months of the year, but the overt behavior of most
group members is constantly restricted by the presence of
dominant group members. An action is or is not permissible
depending on who is watching and who might lend support.
Success requires that a monkey know and integrate the status
of all group members present, and their alliances and antag-
onisms toward him and among each other. Thus, a primate,
in his relationship with a partner, is able to use a third animal. A female can provoke an attack from a male against an
opponent who ranks above her by the trick of presenting to

Fig. 6.1. An adult female chimpanzee using a stick for scratching. (Delta Primate Research Center.)

the male while threatening the opponent. Subordinate male macaques approaching dominant males often take along a young infant to inhibit the aggression of the other. Whereas wild monkeys use no technical tools to exploit their habitat, they manifest analogous schemes in their social behavior.

According to Chance, "tool-using" in the social context may have predisposed the ancestors of man to develop technical tools. This speculation has an interesting secondary implication. If the primate ability of predicting combined effects was indeed transferred from the social to the technical context, then this ability was at the same time freed of the ancient compulsions inherent in social behavior, such as aggression and sex. The handling of sticks or wheels is not loaded with the emotions that go with the handling of social partners, and progress in technology could thus be much faster than progress in social behavior. This disparity is a major problem of modern man.

This leads us back to the social skills of nonhuman primates. Skilfull behavior in the social field require that the actor be capable of adapting his own emotional behavior to the situation—which now and then means suppressing it. Without this ability, predicting and combining would be useless. It seems that some primates can indeed suppress certain behaviors even against strong motivations to act. The female anubis baboons that were transferred to a hamadryas troop offer an example. They easily learned that the aggression of the herding hamadryas male could be avoided by staying close to him, but this meant they had to suppress their strong motivation to flee. Most anubis females succeeded in doing so for many hours on end, although sudden compulsive escapes occurred even long after the new social role had been learned.

The ability to predict combined effects and to control one's own behavior may thus be among the primate predispositions for human adaptations. These abilities would have to be generalized and transferable in order to serve as a basis for culture. So far, field studies have been crude in their approach to these abstract aspects of behavior. We do not know how much combination and self-control are worked into the social behavior of primates, and we have not studied their group foraging with special attention to these abilities. This, perhaps, is why we are often left with the impression that behavior and social organization are no more refined among primates than among many other mammals. We may have looked too much at the units of behavior and not enough at its organization. Experimental results from laboratories should receive much more attention from field workers than in the past.

SEXUAL DIFFERENTIATION AND GROUP LIFE

If we now shift to the more ancient levels of social roles and the style of group life, comparisons become more easy and traits more common in primates and man.

The primate male, in general, is more aggressive and more dominant than the female. He is more likely to leave the group and to migrate. This can probably be said of most human societies too. Within the family group, the human male seeks his activities farther from the home base than the female. Whether these are more than superficial similarities is open to question. In general, human cultures seem to push the sexual division of roles much further than nonhuman primate societies.

This differentiation of roles is already significant with primate infants. The play groups of macaques and baboons, for example, generally include more juvenile males than females; in hamadryas play groups the ratio is about eight to one. While the males are out playing, the females often remain with the female adults of their family group. Sociographic analyses further show that male juveniles interact in larger groups than females, who mostly associate with one partner only. Preliminary data based on the same methods reveal a similar pattern in human children.

One promising way of comparing human and primate behavior it to ask: What is the total array of social tendencies evolved in the primate order? And which of these tendencies have passed into the human heritage? The characteristics most interesting to compare are not the simple motor acts and communicative signals cherished by ethologists, but the higher level of behavioral sets such as submission or social exclusion. These sets make use of the communicative acts, but they are highly independent of their particular form and have their own taxonomic distribution. The style of primate group life may exemplify this level of comparison.

The style of primate groups apparently varies along a main gradient, with the macaque-baboon style and the chimpanzee style as the extremes. Baboon and macaque societies are typically characterized by intense dominance. Individuals tend to assert exclusive access to a particular partner. On the group level, this tendency is paralleled by a strong differentiation between group members and outsiders, with avoidance or antagonism between neighboring groups. Thus, the groups are typically closed and live separately, with the hamadryas and the geladas as the only major exceptions. It is still unclear whether dominance and exclusiveness are causally connected parts of the same syndrome. Territoriality, it seems, is not closely associated with either of them.

The group style of chimpanzees and (largely) of gorillas is marked by a low intensity of dominance. Exclusive claims for partners are absent among adults, so that an inferior male can copulate with a female in full view of the dominant male. These great apes do not noticeably discriminate against outsiders. Their society is open, and its members, instead of living as a closed pack of "ins" among slightly inimical neighbors, are socially and spatially mobile. Fights between entire groups have been seen among macaques, baboons, and langurs, but not among the large apes (Fig. 6.2).

One would expect that man, a close relative of the apes, would approach their social style rather than that of baboons. But the overall impression suggests the opposite. Man's latent or overt inclinations for dominance hierarchies, closed groups, and discrimination against outsiders suggest that he approached the baboon type of society, at least at one stage of his evolution. In many respects the hamadryas baboon's society of closed but coordinated family units is a better model of human social structure than that of the chimps. While man seems on the way toward open societies, his rigid social attitudes are often transferred to larger groups, to the level of professions, religious, nations, and races, and they continue to flare up on the level of small groups. The British primatologist Vernon Reynolds, drawing attention to this phylogenetic incongruence, suggests ecological explanations. He points out that such immobile investments of labor as crop fields, stores, cattle, and houses must have worked in favor of territorial behavior, closed food-sharing units, and hierarchies based on exclusive possessions. Regardless of whether man ever passed through a chimp-like stage, it is obvious that behavior sets of this type have a peculiar taxonomic distribution. They are not confined to a closely related group of species but they emerge here and there without apparent systematic continuity. Territorial behavior, dominance, and responses of social exclusion appear to be general vertebrate potentials. They seem to emerge in very similar forms, by evolution or modification, wherever a species is faced with appropriate ecological conditions.

CONCLUSION

The speculative and deductive character of this text may at times have disappointed the reader as much as it frustrated the author. While thinking and writing, I came to see several reasons why primate field studies have so far failed to present conclusive insights into the ecological functions of societies.

First, social behavior is of two kinds. One kind consists of the behaviors that establish and constantly reestablish the

Fig. 6.2. Latent antagonism between baboon groups may flare up even when the groups normally tolerate each other's proximity: A band of hamadryas baboons (foreground) drives another band (rear, on slope) from an artificial feeding ground.

society. These are mating, nursing, fighting, playing, social grooming, and other associative interactions. These behaviors occur mostly at resting places; they are conspicuous and have therefore been well studied. The second category shows the society in function, in its concerted interaction with the habitat. Spatial arrangements in social foraging and traveling, decisions on travel routes, and communication about food sites belong to this class. Its manifestations are subtle and inconspicuous, consisting of a short glance or of a male's sitting down instead of walking on. Studying them is difficult and has therefore been neglected, although it is this class of behavior that passes or fails in the ecological test, not the noisy fights. With respect to the present theme, field students have generally looked at the wrong side of the picture in studying a society's internal physiology instead of its ecological functions. Let us be fair—not all investigators who went to the field intended to study adaptive functions.

Second, it seems that future field studies pertinent to our subject must shift from the easily observable motor patterns of the individual to the higher level of behavioral sets and strategies. These are more relevant to survival than the particular form of a threat or a digging movement. We have seen that the strong point of primate adaptation does not lie in the motor skills of the individual, but in the way things are done in groups.

Third, there is a need for increased experimental research, both in the field and in laboratories, on the range of modification in response to varied environments. It is not enough to describe one variant of a species' social organization that occurs under "natural" conditions. We should investigate the modification potential of a species to its very limits, that is, to the point where the changes induced by the environment are no longer adaptive and homeostatic but lead to breakdown. I can hardly imagine a more urgent research task than to gather such knowledge about man. Insights into the tolerance limits of primates could help us in defining our own.

Suggested Further Reading

References cited in this volume are marked by an asterisk.

GENERAL WORKS

Altmann, S. A., *Social communication among primates*. Chicago: University of Chicago Press, 1967. 392 pp. A conference volume including original papers from the field and the laboratory on various categories of social behavior and communication, and on their physiological bases.

Chance, M. R. A., and C. Jolly. *Social groups of monkeys, apes and men*. London: Jonathan Cape, 1970. 224 pp. A description of social structures and processes of socialization among the better known primate species, and a discussion of the underlying behavioral mechanisms.

*Crook, J. H. The socio-ecology of primates. In J. H. Crook, ed. *Social behavior in birds and mammals*. New York: Academic Press, 1970, pp. 103–166. Crook here reviews, contributes and discusses hypotheses and speculations on the adaptive values of primate social structure, especially of group size, group composition, and socialization.

DeVore, I., ed. *Primate behavior—Field studies of monkeys and apes*. New York: Holt, Rinehart and Winston, 1965. 654 pp. This is the first collective volume on primate behavior in the wild. It contains descriptive accounts, mostly by the investigators. The theoretical background is based on the thinking of Sherwood L. Washburn.

Jay, P. C., ed. *Primates—Studies in adaptation and variability*. New York: Holt, Rinehart and Winston, 1968. 529 pp. A second collection of field reports, following the one edited by DeVore. The emphasis here has shifted from the description of species to the treatment of topics. Adaptive functions are approached by comparing populations in different habitats.

Napier, J. R., and P. H. Napier. *A handbook of living primates*. New York: Academic Press, 1967. 456 pp. A concentrated catalogue of facts, classified by genera, including morphology, range, ecology, reproduction, behavior, and other aspects of primate biology.

Schrier, A. M., H. F. Harlow, and F. Stollnitz, eds. *Behavior of nonhuman primates.* 2 vols. New York: Academic Press, 1965. As a counterpart of the field volumes, this collection depicts the performance of primates in laboratories. The two volumes include reviews of learning, investigation, perception, and affectional behavior. Many of these articles focus on individual development.

Thorpe, W. H. *Learning and instinct in animals.* 2nd ed. London: Methuen, 1969. 558 pp.

SPECIFIC STUDIES

*Altmann, S. A., and I. Altmann. Baboon ecology. *Bibliotheca primatologica* 12: 1–220 (1970). The most thorough study of primate ecology available, this monograph is based on field research on the yellow baboon (Papio cynocephalus), but also reviews data from other baboon species.

*Carpenter, C. R. A field study of the behavior and social relations of Howling Monkeys. In C. R. Carpenter, *Naturalistic behavior of nonhuman primates.* University Park, Pa.: Pennsylvania State University Press, 1964, pp. 3–92.

Crook, J. H. Gelada baboon herd structure and movement. *Symp. Zool. Soc. London* 18: 237–258 (1966). The first and only original report on geladas in the wild establishes their organization in one-male groups.

*Crook, J, H., and P. Aldrich-Blake. Ecological and behavioral contrasts between sympatric ground-dwelling primates in Ethiopia. *Folia primatologica* 8: 192–227 (1968). A field study on a location where anubis and gelada baboons occur together but nevertheless differ in their habits.

*Ellefson, J. O. Territorial behavior in the common white-handed gibbon, Hylobates lar. In P. C. Jay, ed., *Primates—Studies in adaptation and variability.* New York: Holt, Rinehart and Winston, 1968, pp. 180–199. Confirming Carpenter's classical studies, Ellefson describes the aggressive encounters of neighboring gibbon pairs on their common borders. The paper includes a discussion of the phylogeny and adaptive significance of the trait.

*Gartlan, J. S., and C. K. Brain. Ecology and social variability in Cercopithecus aethiops and C. mitis. In P. C. Jay, ed., *Primates—Studies in adaptation and variability.* New York: Holt, Rinehart and Winston, 1968, pp. 253–292.

*Hall, K. R. L. Behaviour and ecology of the wild patas monkey, Erythrocebus patas, in Uganda. *Journal of Zoology, London* 148: 15–87 (1965). The first and only comprehensive study of the social and ecological behavior of the patas.

*Kawai, M. Newly acquired precultural behavior of the natural troop of Japanese monkeys on Koshima islet. *Primates* 6: 1–30 (1965). A detailed report on the beginnings of a primate tradition (food-washing).

Kummer, H. *Social organization of hamadryas baboons.* Chicago: University of Chicago Press, 1968. 189 pp. The first and only comprehensive report on hamadryas baboons in the wild, based on field research in Ethiopia. The monograph covers all the hamadryas material presented in this volume except for the unpublished data on the transplantation of anubis females and on inhibition among males.

van Lawick-Goodall, J. The behavior of free-living chimpanzees in the Gombe Stream Reserve. *Animal Behaviour Monographs* 1 (3): 161–311 (1968). A full and well-illustrated report focusing on social behavior at a feeding site. The special merit of the study is its duration, which permits us to follow the social development of individuals.

Mason, W. A. Use of space by Callicebus groups. In P. C. Jay, ed., *Primates—Studies in adaptation and variability.* New York: Holt, Rinehart and Winston, 1968, pp. 200–216. A concise description of the only clear case of primate territorial behavior besides that of the gibbon.

*Menzel, E. W. Responsiveness to objects in free-ranging Japanese monkeys. *Behaviour* 26: 130–149 (1966).

Nagel, U. Social organization in a baboon hybrid zone. *Third International Congress of Primatology,* in press.

Reynolds, V. Open groups in hominid evolution. *Man* 1: 441–452 (1966). A student of chimpanzees in the wild visualizes the phylogeny of social structures in man on the basis of paleontological evidence and the social behavior of the living apes. An excellent basis for discussion.

*Ripley, S. Intertroop encounters among Ceylon Gray Langurs (Presbytis entellus). In S. A. Altmann, ed., *Social communication among primates.* Chicago: University of Chicago Press, 1967, pp. 237–254.

Sade, D. S. Inhibition of son-mother mating among free-ranging rhesus monkeys. *Science and Psychoanalysis* 12: 18–38 (1968). This fascinating study of the multiple causation of the low frequency of son-mother matings can also serve as an introduction to the problems of causal investigation by observation alone.

Schaller, G. B. *The mountain gorilla—Ecology and behavior*. Chicago: University of Chicago Press, 1963. 431 pp. The first and only comprehensive report on free-living gorillas; the author refrains from theorizing but provides a wealth of solid facts on a species that had long been considered too dangerous for study.

Struhsacker, T. T. Correlates of ecology and social organization among African cercopithecines. *Folia primatologica* 11: 80–118 (1969). A reevaluation of recent speculations in the light of the author's own studies of forest-living primates.

————. Observations on the behaviour and ecology of the patas monkey (Erythrocebus patas) in the Waza Reserve, Cameroon. *Journal of Zoology, London* 161: 49–63 (1970).

*Yoshiba, K. Local and intertroop variability in ecology and social behavior of common Indian langurs. In P. C. Jay, ed., *Primates— Studies in adaptation and variability*. New York: Holt, Rinehart and Winston, 1968, pp. 217–242. A comparative report on the species which, according to present knowledge, displays the greatest intraspecific variability of social behavior among primates.

FILMS

Kummer, H., W. Götz, and W. Angst. Adaptation of female anubis baboons to the social system of hamadryas baboons. 30 minutes. Information from: H. Kummer, University of Zurich, Birchstr. 95, 8050 Zurich, Switzerland.

Marler, P., and H. van Lawick, Wild Chimpanzees. 40 minutes. On rent and sale from: The Rockefeller University Film Service, New York, New York 10021.

Struhsacker, T. T. Behavior and Ecology of Vervet Monkeys, Cercopithecus aethiops. 40 minutes. On rent and sale from: The Rockefeller University Film Service, New York, New York 10021.

Index